A Practical Guide for Feeding Captive Reptiles

A Practical Guide for Feeding Captive Reptiles

Fredric L. Frye
B.S., D.V.M., M.S.
Fellow, Royal Society of Medicine

KRIEGER PUBLISHING COMPANY
Malabar, Florida
1993

Original Edition 1991
Reissue 1993 with corrections

Printed and Published by
KRIEGER PUBLISHING COMPANY
KRIEGER DRIVE
MALABAR, FLORIDA 32950

Copyright © 1991 by Krieger Publishing Company

All rights reserved. No part of this book may be reproduced in any form or by any means, electronic or mechanical, including information storage and retrieval systems without permission in writing by the publisher.
No liability is assumed with respect to the use of the information contained herein.
Printed in the United States of America

Library of Congress Cataloging-in-Publication Data

Frye, Fredric L.
    Practical guide for feeding captive reptiles/Fredric L. Frye—
Original ed.
      p. cm.
    Includes bibliographical references and index.
    ISBN 0-89464-840-3 (alk. paper)
    1. Captive reptiles—Feeding and feeds. I. Title.
SF515.F79 1991
639.3"9—dc20
                                                                                          90-49750
                                                                                             CIP

10   9   8   7   6   5   4   3   2

Contents

Acknowledgments viii
Foreword xi

Chapter 1: Basic Nutrition for Captive Reptiles

Introduction 1
Basic Nutritional Considerations for Captive
　　Reptiles 2
Preferred Diets 4
Water Requirements, Dehydration, and Gout 41
Salt Metabolism and Requirements 43
Frequency of Feeding 44

Chapter 2: Nutrition-Related Illness and Its Treatment

Introduction 47
Anorexia, Inanition and Starvation 48
Forced Feeding 50
Mineral Deficiencies, Imbalances, and Excesses ... 50
Vitamins and Vitamin Deficiencies 58
Hypoglycemia in Captive Crocodilians 67
Constipation, Obstipation, and Gastrointestinal
　　Blockages 68
Vomiting 70
Diarrhea 71
Bloating (Tympany) 72
Articular and Periarticular Pseudogout 72
Toxic Plant Poisoning 72
Special Nutrition-Related Behaviors 79

Chapter 3: Sources of Reptile Food

    Introduction 83
    Hydroponic Cultivation of Grasses 88
    Bean, Pea, Lentil and Seed Sprout Culture 89
    High-Fiber Diet for Herbivorous Reptiles 91
    Meat Diets: Notes on the Culture of Prey Species .. 92
    Special Recipes for Aquatic Turtles 108
    How to Prepare, Store, and Use Frozen Food 109

Appendix A: Directory of Herpetological Societies of the United States 113

Appendix B: Glossary of Terms 123

Appendix C: References 129

Appendix D: Species List Cross-Referenced by Common Name 139

Appendix E: Species List Cross-Referenced by Scientific Name 151

Indexes

    Index of Taxonomic Names 163
    General Subject Index 169

LIST OF TABLES AND CHARTS

Food Preferences for Selected Snake Species	5
Food Preferences for Selected Lizard Species	20
Food Preferences for Selected Turtles, Terrapins, and Tortoises	31
Food Preferences for Most Crocodilians	40
Food Preferences for the Tuatara	41
Mineral Content of Mice	52
Toxic Plants	74
Nutritious Plants	84
Food Values	86
Chemical Composition of Silk Moth Larvae	97
Chemical Composition of Calcium-Enriched Silk Moth Larvae	97
Chemical Composition of Meal Worms	102
Culture Medium Formula	104
Culture Technique for Wax Moth Larvae	106
Some Sources of Living and Frozen Reptile Food	107
Commercial Food Products for Turtles	110

Acknowledgments

A well-earned debt of thanks is due to my wife Brucye who cheerfully read several drafts of the manuscript for this book, and to Mary Roberts and Edna Perkins of the editorial staff, and to Marie Bowles, Production Manager of the Robert E. Krieger Publishing Company, for their perspicacity, enthusiasm, and good humor in editing the manuscript and producing this guide. Each made suggestions that enhanced the "user friendliness" of the final draft immeasurably.

DEDICATION

To Brucye, Lorraine, Erik, Bice, Noah and Ian.

To the memory of my parents, who showed so much forbearance during my childhood when they allowed our home to be populated by many scaled, furred and feathered creatures.

To our long-suffering neighbors, who never knew what beastie would show up next in their garden or on their doorstep.

BOOKS BY THE AUTHOR

Husbandry, Medicine and Surgery in Captive Reptiles

Biomedical and Surgical Aspects of Captive Reptile Husbandry

Phyllis, Phallus, Genghis Cohen & Other Creatures I Have Known

First Aid for Your Dog

First Aid for Your Cat

Schnauzers, A Complete Owner's Manual

Mutts, A Complete Owner's Manual (simultaneously published in the United Kingdom as Mongrels, A Complete Owner's Manual)

Biomedical and Surgical Aspects of Captive Reptile Husbandry, 2nd Ed.

Captive Invertebrates: A Guide to Their Biology and Husbandry

A Practical Guide for Feeding Captive Reptiles

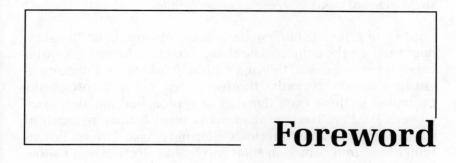

Foreword

The popularity of reptiles as pets has never been greater. This trend has been enhanced over the years by books such as Carl Kauffield's *Snakes and Snake Hunting* and *Snakes: The Keeper and the Kept*. Pet stores have offered an increasing variety of reptiles and supplies, granting ready access to the hobbyist. Reptiles also fit into the busy lifestyles of the nineteen-nineties because they require little time to maintain. The popularity of keeping reptiles has resulted in the formation of almost 80 regional herpetological societies in the United States and the number of advertisements for reptile breeders and sellers or herpetological books, equipment, and supplies has never been greater. The readers of this book, being reptile enthusiasts, no doubt already appreciate their appeal.

One disadvantage of reptile popularity is the problem of providing adequate nutrition. Not only are complete and balanced reptile "chows" often unavailable commercially, but the nutritional requirements of reptiles are often poorly understood. The dilemma is amplified by the large number of diverse species kept as pets, each with its unique requirements and food preferences. Therefore providing wholesome, balanced diets that are readily accepted is a complex and difficult problem, and information on the subject has been generally unavailable to the average person. That was until Doctor Frye wrote this book.

Fredric L. Frye, B.S., D.V.M., M.S., F.R.S.M., pioneered herpetological medicine in the United States. His three textbooks, *Husbandry, Medicine, and Surgery in Captive Reptiles*, and first and second editions of *Biomedical and Surgical Aspects of*

Captive Reptile Husbandry have each, in turn, been the standard texts on the subject. The sheer mass of information, number of reference cases, thorough bibliographies, and documentation through literally thousands of color photographs presented in these texts remains unapproached, let alone surpassed. Dr. Frye's accomplishments were further recognized when he received the American Veterinary Association's *Practitioner Research Award* in 1969 and by his election as a Fellow by the Royal Society of Medicine in 1989. I am proud to call him my colleague and friend.

Dr. Frye has now addressed the problem of helping professional and amateur herpetologists provide their reptiles with proper diets. Moreover, he has done so in a very practical manner. Quite simply, he lists preferred diets for virtually every species of reptile that might find its way into a collection; nutrition-related illnesses and their causes and treatment; and sources of reptile foods. The reptile keeper will find an extraordinary amount of useful information in a very concise format.

The preferred diets are supplemented with descriptions of water and salt metabolism, and suggestions on feeding frequency. The discussion of nutrition-related illness includes not only vitamin and mineral imbalances, but also lack of appetite, forced feeding, constipation, vomiting, diarrhea, gastrointestinal bloating, cannibalism, and the ingestion of shed skin, eggs, stones, and excrement. The section on sources of reptile foods includes advice on the production of sprouts and hydroponically grown grasses, as well as information on how to raise earthworms, mealworms, crickets, fruit flies, and silk moth larvae. Names and addresses of vendors of live and frozen rodents, insects, and aquatic turtle chows are provided. Lists of both toxic and nutritious plants will prove especially useful.

A glossary is provided to clarify any technical terms, along with indexes of species names. Doctor Frye had the forethought to list these by both their common English names and also by their scientific Latin names, so knowing one will enable the reader to quickly find the other. The bibliography is sufficiently comprehensive to permit the serious student to pursue the subject at greater depth. Further enhancing the value of this text is an appendix listing regional herpetological societies by state.

This will permit the reader to contact fellow enthusiasts in his or her geographical area.

A Practical Guide for Feeding Captive Reptiles is exactly what its title claims to be. It will go a long way toward making this complex topic understandable, and is full of useful information for every reptile keeper.

Stephen L. Barten, D.V.M.
Vernon Hills Animal Hospital
Mundelein, IL 60060

1

Basic Nutrition for Captive Reptiles

INTRODUCTION

But if you tame me, then we shall need each other. To me, you will be unique in all the world. To you, I shall be unique in all the world. . . . You become responsible, forever, for what you have tamed.

Antoine de Saint Exupéry
(1900–1944)

These words penned over fifty years ago by Saint Exupéry in "The Little Prince" are as true today as they were then. Whatever species of animals are selected for captive care, it is essential to remember that the mere fact of captivity means that the animals are entirely dependent upon their keeper for all of their needs for adequate shelter, food, water, and health maintenance.

Because of their widely diverse dietary preferences and modes of eating, the Class Reptilia is an interesting and challenging group of animals for the comparative zoologist in general, and the animal nutritionist specifically. Some reptiles are strictly herbivorous; some are carnivorous; yet others are omnivorous. Some restrict their diet to only a single item of food; others will accept a much more varied diet.

Usually for the beginner, the wisest choice is to start with indigenous species whose dietary needs can best be met with the least amount of effort and expense. In climates with severe winter weather, obtaining natural prey or fodder may be impossible. It is sad to see hobbyists whose first reptiles require highly specialized diets not able to feed their animals properly for most of the year. Sometimes the right food items are available only from commercial sources at inflated costs. In this guide, ready sources or alternatives for many dietary items are discussed in detail. In the case of invertebrate prey, practical methods for their culture are provided.

Whenever possible, the begining herpetologist should select a reptile which is native to his or her own general geographic area. For most of North America and Europe, many small temperate life zone natricine snakes, insectivorous lizards, and terrestrial or semiaquatic chelonians are locally available. Garter snakes and small iguanid, gerrhonotid, scincid, gekkonid, or agamid lizards are ideal for the launching of an exciting hobby or nascent career in herpetology.

The sophisticated and thoughtful amateur herpetologist gathers no more animals from the wild than he or she actually requires, and disturbs the collecting site as little as possible. Local game laws must be obeyed and the necessary permits must be obtained prior to actually taking the specimens. In some localities, permits are required to possess some species of native and non-native or "exotic" reptiles.

Throughout this manual, certain diagnostic methods and treatments will be mentioned. Although the purpose of this guide is to provide the correct diets for a variety of captive reptiles, it is not intended as a do-it-yourself veterinary manual for non-veterinarians. Rather, the herpetologist or hobbyist, recognizing a medical problem that may be described in this publication, is encouraged to seek the professional services of a veterinarian experienced in treating non-domestic species of animals.

BASIC NUTRITIONAL CONSIDERATIONS FOR CAPTIVE REPTILES

Introduction

Before any reptiles are brought home or to the collection, it serves the amateur herpetolologist well to become familiar with the habitat and nutritional requirements for each and every animal that he or she plans to keep in captivity. As noted earlier, captive diets for some of these animals can be particularly difficult to provide during some of the winter and hot dry summer months.

Although all snakes and crocodilians are carnivorous, many lizards and chelonians (turtles, terrapins, and tortoises) prefer a more varied diet; they may be carnivorous, herbivorous, or somewhere in between and are termed omnivorous. The omnivorous species accept food items of animal and vegetable origin. Furthermore, some chelonians and fruit and leaf-eating lizards actually change their dietary preferences as they grow to maturity. During their juvenile phase when body size and internal organs expand greatly, they require a diet rich in animal and plant protein so that growth can continue efficiently and in a timely fashion. Once they approach adult size and sexual maturity, growth slows and the need for relatively large amounts of animal protein diminishes and these animals actively select a more plant-based diet. A secondary benefit of this dietary change is that less energy must be expended in foraging for edible plants because they cannot escape predation as do many animals who are capable of rapid movement.

The concept of parallel evolution also should be understood: animals of vastly different genetic heritage living in widely separated parts of the world, but sharing the same ecological niche, often evolve in such a fashion that they look and, sometimes behave almost identically. Some of the best examples of this form of parallel evolution can be found in fishes, reptiles, birds, and mammals, especially marsupials. With respect to reptiles, two lizards can serve as examples: the common new world iguana and the southeast Asian water dragon. Each of these robust lizards lives in a rain forest habitat and each tends to be at least partially tree-dwelling. They are of very similar body shape and size when adult. Their skin color is nearly identical and each carries a crest of spines along its back. Although each of these lizards evolved a half a world apart, they have come to look so much alike that some herpetologists have mistaken each for the other, especially when examining young of each species. However, when one considers the diets that each animal consumes, this parallelism tends to diverge from a common line: iguanas prefer a diet of leaves and fruit; water dragons prefer fish, invertebrates, and small mammals and birds. Iguanas will also eat animals and water dragons will accept fruit, but the staple diets of each are well defined and quite different for iguanas usually refuse fish and water dragons usually will not eat leaves. Unless one knows one lizard from another, it is not likely that an adequate diet will be provided.

PREFERRED DIETS

Snakes

Snakes are carnivorous creatures; depending upon their dietary preferences their natural prey encompasses mammals, birds, other reptiles, amphibians, fish, or invertebrates such as worms, slugs, snails, or insects. Highly specialized snakes feed *entirely* upon eggs or other snakes, amphibian larvae, slugs, crayfish, termites, and even scorpions. Therefore, it is vital to be familiar with the snakes' taxonomy in order to be able to recommend an appropriate diet. Some snake species such as the garter snakes, indigo snakes, and some racers accept a very wide variety of warm-blooded and cold-blooded prey. Some snakes alter their dietary preferences as they grow. Eating invertebrates as juveniles, they change to a diet of lizards and or small rodents as they grow larger.

Because they swallow their prey *in toto*, snakes consume 100% of each meal. An exception to this generality is the egg-eating snake, *Dasypeltis scabra*, which, after ingesting an intact egg many times larger than its head's dimensions would suggest being possible, neatly slits the egg in its esophagus. After swallowing the egg's contents, the snake regurgitates the shell fragments. The digestibility of each meal is, of course, variable. For instance, most arachnids and insects such as mealworms and crickets are encased in chitinous exoskeletons or cuticles that impede total digestion of the softer body parts; these arthropods may yield less of their potential energy content than softer-bodied insects such as moth larvae, fly maggots, etc. However, many reptiles are rather particular in what they will accept as

Table 1
Food Preferences for Selected Snake Species

Snake Species	SM	B	OS	L	E	F/T/S	F	I/A	W/S
Aesculapian snake *Elaphe l. longissimus*	X	O							
African beaked snake *Rhamphiophis multimaculatus*	X			O					
Argentine green speckled snake *Leimadophis poecilogyrus*							small live fish		
African house snake *Lamprophis fulginosus*	X	X		O	O				
Anaconda *Eunectes murinus, Eunectes notaeus*	X	X	X	O	O	O	X	X	
Asian rat snake (mangrove rat snake) *Gonyosoma* sp.	X	X							
Australian tree snake *Dendrelaphis punctularus*	X	X		O		X			
Bandy-bandy snake *Vermicella annulata*			X						

X = usual food items; O = occasionally eaten; SM = small mammals; B = birds; OS = other snakes; L = lizards; E = eggs; F/T/S = frogs, toads, salamanders; I/A = insects and arachnids; F = fish; W/S = worms and slugs.

Table 1 (Continued)

Snake Species	SM	B	OS	L	E	F/T/S	F	I/A	W/S
Bird snake Thelotornis kirtlandii	X	X		X		X			
Black-headed Tantilla sp.								X	O insect larvae and arachnids
Black-striped Coniophanes imperalis	X		X	X		X			
Black swamp snake Seminatrix sp.						O	X		earthworms
Black tree Thasops jacksoni	X		X		X		Frogs		
Blind Rhamphotyphlops etc.				O				X	
Blunt-headed tree snake Imantodes cenchoa				X					
Boa/python Aspidites, Boa, Calabaria, Candoia, Charina, Chondropython, Corallus, Exiloboa, Epicrates, Eryx, Liasis, Lichanura, Morelia, Python Trachyboa, Tropidophis, Ungaliophis, etc.	X	X		X	O	O in some dwarf boas	O		
Boomslang Dispholidus typus	X	X	O	X	O				

Basic Nutrition for Captive Reptiles   7

Species	SM	B	OS	L	F/T/S	E	I/A	F	W/S
Brown tree *Boiga irregularis*	X	X				X	O	O	
Bull snake *Pituophis melanoleucus*	X	X							
Bushmaster *Lachesis muta muta*	X	X			frogs				
Cantil *Agkistrodon bilineatus*	X	O	X	O		X	X		
Cat-eyed, North American *Leptodeira septentrionalis;*				X	X				
Cat-eyed, South American *Boiga sp.*	X			X	X				
Coachwhip *Masticophis sp.*	X	X	O	X	O	O			O
Cobra (Exc. King Cobra) *Aspidelaps sp.; Boulengerina sp.; Hemachatus haemachatus; Naja sp.; Pseudohaje goldii; Walterinnesia aegyptiae*	X (quail chicks)	O	O						
Copperhead *Agkistrodon contortrix*	X		O	O	O	X	O (cicadas and larvae)		

X = usual food items; O = occasionally eaten; SM = small mammals; B = birds; OS = other snakes; L = lizards; E = eggs; F/T/S = frogs, toads, salamanders; I/A = insects and arachnids; F = fish; W/S = worms and slugs.

Table 1 (Continued)

Snake Species	SM	B	OS	L	E	F/T/S	F	I/A	W/S
Copperhead racer *Elaphe radiata*	X	X							
Coral *Micrurus* sp. *Micruroides* sp.	mouse pups		X	X				X	O
Corn snake *Elaphe guttata*	X	X							
Crawfish or swamp *Regina* sp.						amphibians and their larvae	X	crawfish and aquatic nymphs	
Cribo *Drymarchon corais*	X		X	X		X	X		
Crowned snake, New World *Tantilla* sp.								X	X
Crowned snake, Old World *Coronella girondica*	X		X	X				X as juveniles	
Death adder *Acanthophis antarticus*	X	X		X					
DeKay's *Storeria dekayi*					tiny salamanders	X		X	X
Diadem snake *Spalerosophis diadema*	X	X		X					

Basic Nutrition for Captive Reptiles 9

Species	SM	B	OS	L	E	F/T/S	I/A	F	W/S
Earth snake *(Rough):* Virginia striatula *(Smooth):* Virginia valeriae							X		X
Egg-eating Dasypeltis scabra	O	O			X				
Elephant trunk or karung Acruchordus sp.								X (goldfish)	
False coral snake Anilius scytale	X		X	X		X			
False coral snake Erythrolampus bizona			X	X		X			
False habu Macropisthodon rudis				X		X			
False water cobra; Brazilian smooth snake Cyclagras gigas	X					X		X	
Flat-headed snake Tantilla gracilis							X		X
Fox snake Elaphe vulpina	X	X							
Garter; ribbon Thamnophis sp.	X	O		X	O	X		X	X

X = usual food items; O = occasionally eaten; SM = small mammals; B = birds; OS = other snakes; L = lizards; E = eggs; F/T/S = frogs, toads, salamanders; I/A = insects and arachnids; F = fish; W/S = worms and slugs.

Table 1 (Continued)

Snake Species	SM	B	OS	L	E	F/T/S	F	I/A	W/S
Glossy snake *Arizona elegans*	X		X	X				O	
Gopher, Bull, Pine *Pituophis melanoleucus spp.*	X	X			O				
Green *Opheodrys sp. Liopeltis sp.*		O	O	X		O		X	O
Ground snake *Sonora semiannulata*								X	
Herald snake *Crotaphopeltis botamboeia*						X	O		
Hog-nosed *Heterodon sp.; Leioheterodon madagascariensis; Xenodon sp.*	O		O rarely	O rarely	X rarely	X	O		
Hook-nosed *Glyalopion sp. Ficimia streckeri*								X	
Indigo *Drymarchon sp.*	X	X	X	X	X	X	X	O	O
Keeled rat snake *Zaocys dhumnades*	X	X	O	O		X			
Kingsnake *Lampropeltis sp.*	X	X	X	X	O	O		O	

Basic Nutrition for Captive Reptiles

Species	SM	B	OS	L	E	F/T/S	F	W/S	I/A
King cobra *Ophiophagus*	O		X	X					
Kirtland's *Clonophis kirtlandi*								X	
Krait *Bungarus* sp.			X	X	O				
Leaf-nosed *Phyllorhynchus decurtatus*				X					
Lined *Tropidoclonium lineatum*						X	O	X	
Long-nosed *Rhinocheilus lecontei*				X	X (and their eggs)				
Lyre *Trimorphodon biscutatus*	X	X		X					
Malayan long-glanded coral snake *Maticora bivirgata flaviceps*			X	X					
Mamba *Dendraspis* sp.	X	X	O	X					
Mangrove *Boiga dendrophila*	X	X		X	X				O

X = usual food items; O = occasionally eaten; SM = small mammals; B = birds; OS = other snakes; L = lizards; E = eggs; F/T/S = frogs, toads, salamanders; F = fish; W/S = worms and slugs; I/A = insects and arachnids.

Table 1 (Continued)

Snake Species	SM	B	OS	L	E	F/T/S	F	I/A	W/S
Marine/sea Acalyptophis peronii; Aipysurus sp.; Astrotia stokesii; Emydocephalus sp. Hydrophis; Laticauda sp. Lapemis sp.; Pelamis platurus							X		
Milk snake Lampropeltis triangulum	X	X	X	X					
Mahogany rat snake **"Puffing" snake** Pseutes sp.	X	X		X		O as juveniles			
Mole snake Lampropeltis calligaster rhombomaculata; Pseudaspis cana	X	X	X	X					
Monpellier snake Malpolon sp.	X	X	X						
Mud Farancia abacura					salamanders esp. Amphiuma				
Mussurana Clelia clelia	X		X	O					
Neck-banded Scaphiodontophis annulatus bondurensis				X					

Basic Nutrition for Captive Reptiles

Species	SM	B	OS	L	E	F/T/S	F	W/S	I/A	Notes
Night/cat-eyed *Eridiphas* sp., *Hypsiglena* sp., etc.	X	X		X		X		O	O	
Parrot snake *Leptophis ahaetulla; Chrysolopea* sp.				X						
Patch-nosed snake *Salvadora* sp.				X						and lizard eggs
Pine snake *Pituophis melanoleucus*	X	X		X						
Pipe snake *Cylindrophis rufus*	X		X							
Pit vipers, Asian *Agkistrodon* sp.; *Calloselasma* sp.; *Deinagkistrodon acutus; Trimeresurus* sp.	X	O		X					O	
Pit vipers, New World *Agkistrodon* sp.; *Bothrops* sp.; *Bothriechis* sp.; *Crotalus* sp.; *Lachesis muta; Ophryacus* sp.; *Porthidium* sp.; *Sistrurus* sp.	X	O		X					O O	in some species
Pygmy rattlesnake *Sistrurus miliarius*	X	O		X						

X = usual food items; O = occasionally eaten; SM = small mammals; B = birds; OS = other snakes; L = lizards; E = eggs; F/T/S = frogs, toads, salamanders; I/A = insects and arachnids; F = fish; W/S = worms and slugs.

14 A Practical Guide for Feeding Captive Reptiles

Table 1 (Continued)

Snake Species	SM	B	OS	L	E	F/T/S	F	I/A	W/S
Queen snake *Regina septemvittata*								crayfish	
Racer (New World) *Alsophis sp.; Coluber sp.; Drymobius sp.; Drymoluber sp.*	X	X	X	X	O	O		O	
Racer (Old World) *Haemorrhois sp.; Chironius sp.*	X	X		O				O	
Rainbow *Farancia erythrogramma*						tadpoles			
Rat/Chicken *Elaphe sp.*	X	X			X				
Rattlesnake *Crotalus sp.; Sistrurus sp.*	X	X	X	X	esp. the Massasauga	O		O	
Red-bellied *Storeria occipitomaculata*						X	X	X	X
Red-necked keelback *Rhabdophis subminiatus*					X	X	O		
Ribbon *Thamnophis sp.*	X	O	X	X	X	X		O	X
Ring-necked *Diadophis sp.*			O	O		X		O	X

Species	SM	B	OS	L	E	F/T/S	F	W/S	I/A
Ringed snake, Grass snake *Natrix natrix*						X	X	O	
Sand snake *Psammophis sp.*	O			X					
Scarlet snake *Cemophora coccinea*	X		X	X	X				
Sharp-tailed *Contia tenuis*								slugs	
Shovel-nosed *Chionactes sp.*									X
Short-tailed *Stilosoma extenuatum*			X	X					
Snail-eating *Dipsas indica; Sibon; Tropidodipsas sartori*								snails and slugs	
Speckled snake *Leimodophis sp.*	O			O		X	X		
Sunbeam snake *Xenopeltis unicolor*	X		X	X					
Taipan snake *Oxyuranus scutellatus*	X								

X = usual food items; O = occasionally eaten; SM = small mammals; B = birds; OS = other snakes; L = lizards; E = eggs; F/T/S = frogs, toads, salamanders; F = fish; W/S = worms and slugs; I/A = insects and arachnids.

Table 1 (Continued)

Snake Species	SM	B	OS	L	E	F/T/S	F	I/A	W/S
Tentacled snake *Erpeton tentaculum*							small live fish		
Tiger snake *Notechis scutatus*	X			X		X			
Tropical chicken *Spilotes pullatus*	X	X		X		X			
Vine snakes *Ahaetulla sp.; Oxybelis sp. Uromacer sp.*		O	X	X					
Vipers, misc. *Agkistrodon sp.; Atheris sp.; Azemiops feae; Bitis sp.; Bothriechis sp.; Bothrops sp.; Calloselasma sp.; Causus sp.; Cerastes sp.; Deinagkistrodon sp.; Echis sp.; Eristocophis sp.; Lachesis muta; Porthidium sp.; Trimeresurus sp.; Vipera sp.*	X	X	O	X		O	O		
Water moccasin *Agkistrodon piscivorus*	X	X	X	X		X	X	O esp. as neonates	O
Watersnakes *Nerodia sp.; Natrix sp.*	O		O			X	X	O esp. as neonates	O

Whip snake *Masticophis sp.*	X	X				
Wolf snake *Lycodon sp.; Lycophidion sp.*	O		X		X	
Worm snakes *Typhlops sp.; Leptotyphlops*						termites, worms and grubs
Yellow-lipped or Pine woods *Rhadinaea flavilata*				X	X	

X = usual food items; O = occasionally eaten; SM = small mammals; B = birds; OS = other snakes; L = lizards; E = eggs; F/T/S = frogs, toads, salamanders; F = fish; W/S = worms and slugs; I/A = insects and arachnids.

prey and may refuse to feed on substitute items. Where appropriate, alternative methods for inducing snakes to accept prey items that are not familiar to them are provided. In the case of insectivorous snakes and lizards, the advantages of feeding mulberry silk moth larvae will be discussed.

Boas of many species, pythons, rat snakes, gopher or bull snakes, old world vipers and new world pit vipers usually prefer warm-blooded prey such as appropriate size rodents and birds. Some, particularly juveniles and the dwarf boas and small tree pythons either prefer or will readily accept tree frogs, lizards, and small snakes as well as small birds and mammals.

Garter snakes, ribbon snakes, water snakes and similar species usually will accept fish, frogs, toads, salamanders, earthworms, slugs, and carrion. Some can be taught to accept mice, either live, freshly killed, or frozen and thawed. Others can be induced to accept dead mice if the mice are first smeared with the mucus of fish or frogs, thus making the rodents appear to the snake as the more normally favored (and flavored) fish or frog prey. An occasional instance of cannibalism of conspecific garter snakes has been reported. Rattlesnakes and water moccasins may, if sufficiently hungry, attack and eat their own kind and an instance was reported of two gopher snakes which cannibalized garter snakes (Zaworski, 1990).

Indigo snakes, kingsnakes, and many racers usually will eat warm-blooded as well as cold-blooded prey such as other snakes, lizards, etc. The indigo snakes have a preference for frogs but, when hungry, will eat almost anything resembling food. Some can be trained to accept canned commercial dog or cat food, but the inherent risk of vitamin A and D overdosage must be considered when including these items in the captive diet. This subject will be discussed in greater detail later.

The smaller ring-necked, sharp-tailed, DeKay's, brown and worm snakes and their closely related kin usually confine their diet to small salamanders (particularly *Batrachoceps* sp.), earthworms, and very small snakes and lizards, especially small skinks.

Racers, vine snakes, coachwhips, etc. tend to prefer lizards, although the racers occasionally also will eat mice and the chicks of small ground-nesting birds such as quail. The young of many of these snakes will readily accept large insects such as crickets, praying mantids, katydids, cicadas, wingless grasshoppers, and locusts.

Some snakes, particularly the king cobra, are almost strictly ophiophagous (snake-eating). They often will accept thawed, frozen snakes or even shed snake skins that have been stuffed with dead mice. If snakes are fed to other snakes, it is usually a wise measure to kill and freeze the snake prey first; this will greatly aid in the reduction of prey-associated food-borne parasitic infestations. Many parasitic organisms will not survive the freezing and thawing cycle, but most bacteria and some protozoans will survive and retain their ability to cause infections.

Lizards

Lizards may, depending upon their natural habits and species, confine their diet to animal protein (in the form of other lizards, birds, mammals, carrion, molluscs, gastropods, insects or eggs), vegetable matter (leaves, blossoms, fruits), or a combination of several basic food classes. Like some snakes, many lizards change their dietary preferences as they mature. For example, common iguanas, *Iguana iguana*, when young are fond of insects, as well as leaves, blossoms, and fruit; as adults they prefer fruits and leafy vegetables, but will accept small mammals and eggs if they are offered. As is the case with snakes, many lizard species tend to be highly specialized in their dietary preferences. For instance, some lizards eat only ants or termites. Others will accept only hard-shelled molluscs and gastropods. Some lizards, like snakes, swallow their prey whole. Others customarily tear their food into bite-sized portions; this is the method employed by the larger monitors and some tegus, which can swallow enormous meals in a short period of time. Other lizards chew their food before swallowing it; this is particularly true of the insectivorous species and the impressive caiman lizard, *Dracaena guianensis*, whose diet of clams, snails and other hard-shelled creatures must be thoroughly crushed in order to extract and swallow the nutritious soft body parts.

Horned lizards (horned "toads"), Australian molochs, and some sand lizards dine almost entirely upon ants and/or ter-

Table 2
Food Preferences for Selected Lizard Species

Lizard Species	SM	B	I	E	CM	M/G	F	V
Agama *Agama sp.*	X		X and arachnids					O
Alligator *Gerrhonotus sp. (= Elgaria)*	X	O	X and arachnids	O	X	O		
American anole *Anolis sp.*			X and arachnids					
Armadillo *Cordylus sp.*			X					O fruit
Basilisk *Basiliscus sp.*	X	quail chicks	X					O
Beaded, Mexican *Heloderma horridum*	X	X	O		X	X		X
Bearded *Amphibolurus barbatus*	X	X	X	X	O	O fish	O	O
Brazilian tree lizard *Enhyalius catenatus*			X and arachnids					
Caiman *Dracaena guianensis*						fish	clams/ snails	
Chameleon Old World *Chamaeleo sp, Brooksia sp.*	O*		X and arachnids				snails	

Basic Nutrition for Captive Reptiles   21

Species									
Chuckwalla *Sauromalus sp.*	O		O		O			X	X blossoms
Collared *Crotaphytus sp.*	X	O other lizards	X and arachnids	O boiled	O			X	X blossoms
Curly-tailed *Leiocephalus carinatus*			X and arachnids						
Deaf agamids *Cophotis sp.*			X and arachnids						
Dwarf sand lizard *Eremias grammica*			X						
Earless *Holbrookia sp., Cophosaurus texanus*			X and arachnids						
Flying dragon *Draco volans*			X and arachnids						
Frilled lizard *Chlamydosaurus kingii*			X						
Fringe-toed *Uma sp.*			X						

*especially *C. parsoni* and other giant species.
X = usual food items; O = occasionally eaten; SM = small mammals; B = birds (chicks); I = insects and arachnids; E = eggs; CM = chopped meat(s); M/G = mollusks/gastropods; F = fish; V = vegetables/fruits.

Table 2 (Continued)

Lizard Species	SM	B	I	E	CM	M/G	F	V
Gecko, misc. Gekko, Hemidactylus, Coleonyx, Gehyra, Hemitheconyx, Phelsuma, Chondrodactylus, Phyllodactylus, Tarentola, Gonatodes, Sphaerodactylus, Ptyodactylus sp., etc.	X	O	X and arachnids	O			O	fruit puree (Phelsuma)
Gila "monster" Heloderma suspectum	X	X	O	X			O	
Girdled lizard Zonosaurus sp.			X and arachnids					O
Glass lizard "Sheltopusik" Ophisaurus ventralis	X	X	X and arachnids	O	X	snails		
Helmeted lizard Corythophanes sp.			X and arachnids					
Horned lizard Phrynosoma sp.			ants, termites					
Iguana, common Iguana iguana. Diet varies with age; see text	O	O	X	X	O	O	X	X esp. leafy
Iguana, desert Dipsosaurus dorsalis	O		X	O	O	O	X	X
Iguana, Fiji Brachylophus sp.	O	O	O	O	O	O	X	X

Basic Nutrition for Captive Reptiles 23

Species											
Iguana, ground *Cyclura sp., Conolophus pallidus*	X	O	O	O	O					X	X algae, kelp
Iguana, marine *Amblyrhynchus sp.*								O			fish algae, kelp
Jungle runner *Ameiva ameiva*			X also arachnids and other lizards								
Lacerta, misc. *Lacerta sp.; Gallotia stelini*	X	X quail chicks	X	O	O			O			
Legless *Anniella pulcra*			X and arachnids								
Leopard *Gambelia sp.*			X and arachnids also arachnids and other lizards								
Long-tailed brush *Urosaurus graciosus*			X and arachnids								
Lyre-headed lizard *Lyriocephalus scutatus*			X and arachnids								
Moloch *Moloch horridus*			ants and termites								
Monitor(s), misc. *Varanus sp.* **(except Gray's**, *V. olivaceus (V. grayi)*)	X	X	O	O	X carrion	O	O	X snails	rarely	X	figs

X = usual food items; O = occasionally eaten; SM = small mammals; B = birds (chicks); I = insects and arachnids; E = eggs; CM = chopped meat(s); M/G = mollusks/gastropods; F = fish; V = vegetables/fruits.

Table 2 (Continued)

Lizard Species	SM	B	I	E	CM	M/G	F	V
Mountain lizard Japalura sp.			ants and termites					
Mountain horned lizard Acanthosaurus armata			X and arachnids					
Night Xantusia sp.; Cricosaura typica			ants and termites					
Plated Gerrhosaurus sp.	X	X	X and arachnids	X	O	O	O	
Rock, banded Petrosaurus mearnsi			X and arachnids					
Ruin Podarcis sicula							O	X
Russian gargoyle lizard Phrynocephalus mystaceus			X and arachnids					
Sail-tailed Hydrosaurus amboinensis	X		O	O		X	X	misc. fruits, leaves and seeds
Sand lizards Psammodromus sp.			X and arachnids					

Table 2 (Continued)

Lizard Species	SM	B	I	E	CM	M/G	F	V
Sharp-snouted snake lizard Lialis burtonis	other lizards		X and arachnids					
Sheltopusik Ophisaurus apodus	X	X and arachnids	X and arachnids	X	X	X		
Side-blotched Uta stansburiana			X and arachnids					
Skinks, small Eumeces, Sincella, Neoceps reynoldsi, Lerista sp.			X small arthropods and arachnids			snails		
Skink, misc. (Old World and Australasian) Scincus, Tiliqua, Chalcides, Trachydosaurus, Egernia sp., Corucia zebrata	O	O	O	O	O	O	X	X fruit puree and nectar
Slow "worm" Anguis fragilis	O		X	O	X	O		
Snake lizard Delma fraseri; Chamaesaura anguina			X and arachnids					
Spiny Sceloporus sp.	X		X and arachnids					

X = usual food items; O = occasionally eaten; SM = small mammals; B = birds (chicks); I = insects and arachnids; E = eggs; CM = chopped meat(s); M/G = mollusks/gastropods; F = fish; V = vegetables/fruits.

26 A Practical Guide for Feeding Captive Reptiles

Table 2 (Continued)

Lizard Species	SM	B	I	E	CM	M/G	F	V
Spiny-tailed *Uromastix* sp.	X	O quail chicks	X	O	O		leafy plants, blossoms, and fruit	
Spiny-tailed iguanas *Ctenosaura pectinata; Urocentron* sp.			O			O	leafy plants, blossoms, and fruit	
Sungazer lizard *Cordylus* sp.			X and arachnids				O	X
Swift, fence, plateau, scrub, rose-bellied, bunch grass, mesquite, sagebrush, etc. *Sceloporus* sp.			X and arachnids					
Swifts, South American *Liolaemus* sp.			X and arachnids					
Tegu *Tupinambis* sp.	X	X	X	X	X	X	X	X
Thorny devil lizard *Heteropterex dilatata*			X and arachnids					
Tree or brush lizards *Urosaurus* sp.			X and arachnids					

Basic Nutrition for Captive Reptiles 27

Species							
Water "dragon" *Physignathus leseurii; P. concincinus*	O	O	X fish	X snails	X live fish, earthworms		O fruit
Whip-tailed *Cnemidophorus sp.*	O		X termites and tiny grubs				
Worm *Bipes biporus, Blanus,*			X and arachnids				
Zebra-tailed *Callisaurus draconoides*							

X = usual food items; O = occasionally eaten; SM = small mammals; B = birds (chicks); I = insects and arachnids; E = eggs; CM = chopped meat(s); M/G = mollusks/gastropods; F = fish; V = vegetables/fruits.

mites. Some will accept small crickets, mealworms and/or small mulberry silk moth larvae as substitutes. Tiny lizards can be fed the larvae of grain weevils, vestigial-winged or wingless fruit flies (*Drosophila melanogaster*), and small fly larvae.

Night lizards, "worm" lizards (amphisbaenids) which the international taxonomic community has classified separately from the "true" lizards, and the smaller geckos usually limit their fare to termites and/or their eggs. Some also accept small beetle or grain weevil larvae and fly larvae; a number of geckos (particularly tropical day geckos) also eagerly devour fruit puree. These fruit preparations may be freshly prepared or purchased as strained infant diets, as the situation dictates.

The common green iguanas, so often sold as pets by the exotic animal trade, thrive on a diet of moistened pelleted alfalfa rabbit or guinea pig chow, dandelion blossoms and leaves, rose petals, chopped turnip, collard or mustard greens, and tender fig, eugenia, hibiscus and mulberry (*Morus*) leaves, frozen mixed vegetables, tofu soybean cake, occasional crickets, silk moth larvae, baby mouse pups, chopped hard-cooked eggs, or fresh fruit (sliced apples, pears, pitted stone fruits, ripe papaya, grated squash, melons, etc.). *Small* amounts of commercial dog or cat food not exceeding 5% of the total ration can be fed, but the bulk of the diet should come from leafy vegetables and fruit. Commercially prepared pet diets, designed to meet the nutritional requirements of carnivores, contain substantial amounts of vitamins A and D that exceed greatly the requirements (and tolerance) of herbivorous lizards. When fed to excess, dog or cat food may induce characteristic lesions of toxicity in reptiles. (This topic will be discussed in greater detail later in this manual.)

If available in sufficient quantities, some commonly grown houseplants, especially hibiscus, nasturtium, and the ubiquitous wandering Jew, *Zebrina pendula*, are avidly eaten and are highly nutritious for these leaf-eating lizards. Some houseplants are, of course, toxic and must be avoided. (See table of toxic plants). As noted earlier, adult iguanas tend to prefer a more herbivorous diet than juveniles, which are far more carnivorous and prefer insects, small rodents, and birds to vegetables, although they do eat vegetable matter as well. Early in their second year, the juveniles begin their gradual dietary preference change to tender shoots, blossoms, and other plant parts, rounding out their intake with ever-lessening amounts of animal protein. (See Auffenberg, 1982; Hirth, 1963; Iverson, 1982; Loftin and Tyson, 1965; McBee and McBee, 1982; Montgomery, 1978; Rand, 1978; Rand et al., 1990; Sylber, 1988; Troyer, 1982, 1984; and Van Devender, 1982).

Hatchling common iguanas have been observed in the wild actively ingesting fresh feces from older conspecific lizards. Troyer (1982) reported that this behavior was directly related to the acquisition of symbiotic microbes necessary for the fermentative processing of cellulose and other carbohydrates.

Tegus, Gila "monsters," Mexican beaded lizards and some of the larger tropical skinks thrive upon a diet consisting of raw eggs, chopped lean meat, baby mice, Japanese quail chicks, and occasional fruit and/or fruit puree. A balanced vitamin-mineral supplement that is rich in calcium carbonate usually is administered once or twice weekly by mixing an appropriate amount with the soft food or fruit puree.

Monitor lizards, depending upon their size (from the smallest to the great Komodo "dragons," which approach 3 meters in length and almost 110 kilograms), will eat a diet similar to that fed to tegus, and supplemented with larger rodents and/or domestic birds. Some monitors prefer fish, others readily accept carrion. One, Gray's monitor, *Varanus olivaceus* (*V. grayi*), prefers fruit, although it will also accept a carnivorous diet of snails.

Geckos of many species are mostly insectivorous; the larger examples also will eat baby mouse pups and/or quail chicks. Many geckos also should have access to fresh or commercially prepared fruit puree or nectar at least twice weekly; strained infant fruits may be substituted if fresh fruit puree is not available. Many of the larger skinks also will accept fruit puree or nectar, and this material makes the task of mineral supplementation much easier to accomplish.

Fence lizards, many skinks, alligator lizards, "glass-snakes," plated lizards, anoles, old world chameleons, etc. thrive on a captive diet of appropriate size insects, from tiny genetically wingless or vestigial-winged fruit flies to crickets, mulberry silk moth larvae, miscellaneous wild moths, grasshoppers and katydids. Where legally available, the larvae of the European corn borer (*Pyrausta nubialis*) are an easily cultured source of insect prey, but they may not contain an adequate amount of calcium for normal bone growth. Wax "worms" (actually the larvae of moths which feed upon the contents of honey bee hives, especially the wax combs) can be cultured readily. (To protect bees from escaped wax moths, the culture chamber must be furnished with a close-fitting cover made from very small mesh which provides ventilation.) Wild moths may be caught by attracting them to bright light bulbs, especially those of blue color, after sundown. Many lizards will eat centipedes, scorpions, and spiders. Mulberry silk moth (*Bombyx mori*) larvae are raised on an exclusive diet of fresh (or thawed properly frozen) mulberry leaves (see Frye, 1987; Frye and Calvert, 1988, 1989a,b, 1991). The smaller skinks tend to be insectivorous and thrive on a diet of assorted winged and larval insects. The larger Australasian species are mostly vegetarian in their dietary preferences although many will also accept small mammals, eggs, and insect larvae such as mulberry silk moth larvae as a minor portion of their rations.

Highly specialized species such as the Galapagos Island marine iguanas, *Amblyrhyncus* sp., eat only certain varieties of marine algae or kelp. The caiman lizard, *Dracaena guaianensis*, with its powerful jaws and flat-cusped molar-like teeth, is superbly adapted to its diet of hard-shelled molluscs and gastropods.

Lizards with less specialized diets may benefit from a varied diet. Vogel *et al.* (1986) demonstrated in their study that two types of insect prey exerted a complementary effect with respect to at least short-term growth, but that prey diversity, *per se*, may not necesssarily enhance growth. Factors influencing these differences were prey water content, prey energy content per dry weight, and assimilation energy efficiencies.

Chelonians

Chelonians (turtles, tortoises, and terrapins) may be carnivorous, herbivorous, or omnivorous. Many marine turtle species

Table 3
Food Preferences for Selected Turtles, Terrapins, and Tortoises

Chelonian Species	M	F	I/W/S/DF	FL	FR	V
African helmeted turtle *Pelomedusa subrufa*	X	X	X			pond weeds and algae
African side-necked turtles *Pelusios sp.*	X	X	X			pond weeds and algae
Aldabra tortoise Geochelone (Testudo) gigantea (Aldabrachelys elephantina)*	O		O	X	X	X
Alligator snapping turtle *Macrochelys temminckii*	X	X	X			
Argentine side-necked turtles *Phrynops sp.*	X	X	crayfish			pond weeds and algae
Asian box turtles *Cuora sp.*	O	X	X	O	O	O
Big-headed turtle *Platemys macrocephala*	X	X	X			pond weeds and algae
Black pond turtle *Seibenrockiella crassicollis*	X	X	X			pond weeds and algae

*The taxonomic nomenclature regarding this animal is under reconsideration for revision. See Pritchard, (1986)
X = usual food items; O = occasionally eaten; M = misc. meats; F = fish; I/W/S/DF = insects, worms, slugs, snails, small amounts of dog food; FL = flowers; FR = fruits; V = other vegetables.

Table 3 (Continued)

Chelonian Species	M	F	I/W/S/DF	FL	FR	V
Blanding's turtle *Emydoidea blandingi.*	X	X	X	O	X	pond weeds and algae
Bog turtle *Clemmys muhlenbergi*	\multicolumn{6}{l}{Juveniles tend to be more carnivorous than adults.}	
	X	X	X			O
Bolson tortoise *Gopherus (Xerobates) flavomarginata*		O	O	X	X	X
Box turtle *Terrepene sp.*	X	O	X	X	X	X
Brahminy River turtle *Hardella thurgi*	X	X	X	X	X	pond weeds and algae
Burmese mountain tortoise *Geochelone emys*			snails	X	X	X
Chaco tortoise *Chelonoidis chilensis*	O		O	X	X	X
Chicken turtle *Deirochelys reticularia*	X	X	X	O	X	pond weeds and algae
	\multicolumn{6}{l}{Juveniles tend to be more carnivorous than adults.}	
Cooter *Chrysemys sp.*	X	X	X			pond weeds and algae
	\multicolumn{6}{l}{Juveniles tend to be more carnivorous than adults.}	

Basic Nutrition for Captive Reptiles

Species	I/W/S/DF	F	M	FR	FL	V	Other
Desert tortoise *Gopherus (Xerobates) agassizi*	O			O (small amounts of dogfood)		X	X
Diamondback terrapin *Malaclemys terrapin*	O	X (Periwinkle snails)		X		X	X (pond weeds and algae)
Egyptian tortoise *Testudo kleinmanni*				O		X	X
Elongated tortoise *Geochelone elongata*				X		X	X
False map turtle *Graptemys pseudogeographica*	X	X		X		X	X (pond weeds and algae)
Fly River turtle *Carettochelys insculpta*	X	X		X		X	X (pond weeds and algae)
Four-eyed turtle *Sacalia bealei*	O	X		X		X	X (pond weeds and algae)
Galapagos tortoise *Geochelone elephantopus sp.*	O			X		X	X
Gopher tortoise *Gopherus polyphemus*	O			O		X	X
Greek tortoise *Testudo graeca*	O			O		X	X

X = usual food items; O = occasionally eaten; M = misc. meats; F = fish; I/W/S/DF = insects, worms, slugs, snails, small amounts of dog food; FL = flowers; FR = fruits; V = other vegetables.

Table 3 (Continued)

Chelonian Species	M	F	I/W/S/DF	FL	FR	V
Green turtle, marine *Chelonia mydas*						
JUVENILES	molluscs, sponges	X				algae, kelp
ADULTS	sponges	O				algae, kelp
Hawksbill turtle *Eretmochelys imbricata*						
JUVENILES	molluscs, sponges	X				algae, kelp
ADULTS	sponges	O				X
Hermann's tortoise *Testudo hermanni*	O		X	X	X	X
Hingeback tortoise *Kinexys* sp.	O		X	X	X	X
Impressed tortoise *Geochelone impressa*	O		X	X	X	X
Leaf turtles, Asian *Cyclemys* sp.	X	X	X earthworms		algae	
Leatherback turtle *Dermochelys coriacea*						
JUVENILES:	jellyfish, sponges and corals	X				algae, kelp
ADULTS:	Jellyfish,	O				algae, kelp

Basic Nutrition for Captive Reptiles 35

Species							
Leopard tortoise *Geochelone pardalis*	O			O		X	X
Loggerhead turtle *Carretta carretta*		sponges and coral	O				O
Long-necked *Chelodina longicollis*		O	X	X			
Map turtle *Graptemys sp.*	X	X	X	X	pond weeds and algae		
Mata mata turtle *Chelys fimbriata*	O		X	O			
Mud turtle *Kinosternon sp.*	X (live)		X	X	pond weeds and algae		

Juveniles tend to be more carnivorous than adults.

X = usual food items; O = occasionally eaten; M = misc. meats; F = fish; I/W/S/DF = insects, worms, slugs, snails, small amounts of dog food; FL = flowers; FR = fruits; V = other vegetables.

Table 3 (Continued)

Chelonian Species	M	F	I/W/S/DF	FL	FR	V
Muhlenberg's turtle *Clemmys muhlenbergi*	X	X	X			pond weeds and algae
		Juveniles tend to be more carnivorous than adults.				
Musk turtle *Sternotherus sp.*	X	X	X			pond weeds and algae
New Guinea side-neck turtle *Elysea novaguineae*	O	X	X			pond weeds and algae
Olive Ridley *Lepidochelys olivacea*		X	molluscs			
Painted turtle *Chrysemys picta*	X	X	X			pond weeds and algae
		Juveniles tend to be more carnivorous than adults.				
Pancake tortoise *Malachocherus tornieri; M. procteri*				X	X	X
Pond turtle *Clemmys sp.*	X	X	X			pond weeds and algae
		Juveniles tend to be more carnivorous than adults.				
Radiated tortoise *Geochelone radiata*	O		O	X	X	X
Red-eared slider turtle *Trachemys scripta elegans*[+]	X	X	X			pond weeds and algae
		Juveniles tend to be more carnivorous than adults.				

Basic Nutrition for Captive Reptiles 37

Species							
Red-footed tortoise *Geochelone carbonaria*	O			X	X	X	
Ringed sawback turtle *Graptemys oculifera*		X	X	X			pond weeds and algae
Side-necked turtle *Chelodina* sp., etc.		X	X	X			
Snapping turtle, common *Chelydra serpentina*		X	X				
Soft-shelled *Trionyx* sp., etc.		X	X				
South American river turtle *Dermatemys mawei*		X	X	X			pond weeds and algae
Spotted turtle *Clemmys guttata*		X	X	X			pond weeds and algae
Spur-thighed tortoise *Geochelone sulcata*				O	X	X	X
Star tortoise *Geochelone elegans*	O			X	X	X	

†The binomial nomenclature for this turtle was formerly *Pseudemys* or *Chrysemys scripta elegans*. It has now been assigned to the genus *Trachemys*.

X = usual food items; O = occasionally eaten; M = misc. meats; F = fish; I/W/S/DF = insects, worms, slugs, snails, small amounts of dog food; FR = fruits; FL = flowers; V = other vegetables.

Table 3 (Continued)

Chelonian Species	M	F	I/W/S/DF	FL	FR	V
Toad-headed *Phrynops* sp.		X	X			pond weeds and algae
Wood turtle *Clemmys insculpta*	X	O	X		O	O
Yellow-footed tortoise *Geochelone denticulata*	O		X	X	X	X

X = usual food items; O = occasionally eaten; M = misc. meats; F = fish; I/W/S/DF = insects, worms, slugs, snails, small amounts of dog food; FL = flowers; FR = fruits; V = other vegetables.

become less carnivorous as they approach adulthood, but most will continue to eat fish and marine invertebrates. Some sea turtles are virtually herbivorous as sexually mature adults although, as juveniles, they subsisted mainly upon fish, molluscs, coelenterates (particularly pelagic jellyfish), and marine vegetation. Some actively seek out and devour coral.

Fresh water turtles and terrapins usually will eat earthworms, small whole fish (preferably live), mouse pups, snails, carrion, algae, and green leafy vegetation such as pond weeds, *Elodea*, *Anacharis*, watercress, duckweed, Swiss chard, romaine lettuce, etc. Many semiaquatic species such as the sliders and pond turtles tend to become far more herbivorous as they approach adulthood; their dietary requirements for high density nutrients diminish as they mature; consequently, they are able to meet these requirements more easily by eating plant materials. Adults of these species also will eat animal sources of food, if the opportunities are present. Fully aquatic marine turtles feed almost exclusively upon marine invertebrates, especially porifera and coelenterates, as well as fish and aquatic vegetation.

Most tortoises usually thrive on a diet of fresh flower blossoms (particularly those with bright red, orange, yellow or green petals), succulents, cactus pads such as the flat "beaver-tail" variety, (*Opuntia*), alfalfa leaves and flowers, or dried alfalfa in the form of commercial pelleted guinea pig or rabbit chows, clover, rose petals, nasturtium and dandelion flowers and leaves, fresh corn (maize) and mixed fruits, squashes (particularly acorn, butternut, banana, and Hubbard), and melons. Fresh figs are an excellent source of highly concentrated carbohydrate(s) given prior to hibernation. Many tortoises will avidly hunt for and eat the common brown house snail, *Helix aspersa*, eating the shell as well as the softer body parts. A *small* amount of commercial dog, cat, or primate chow usually is welcome, but must not be fed to excess because of its high vitamin D content that can induce severe mineralization of soft tissues and its high level of preformed vitamin A which may be toxic for many reptiles. As a general guide, if any of these mammalian chows are fed, they should not exceed 5% of the total ration.

Crocodilians

Crocodilians are almost strictly carnivorous. While most crocodilians will accept a variety of fish, amphibians, reptiles, birds, and mammals, others like the long-snouted gavial (gharial) and false gavials are almost exclusively fish-eating. Their elongated and very narrow jaws, studded with sharp teeth, are well adapted. It is essential that some source of natural whole bone tissue is included with the flesh, rather than merely feeding boneless chunks of meat and/or fish fillets. Some vegetable matter is also ingested by most crocodilians, almost always as the result of incidental swallowing of aquatic grass or the gastrointestinal ingesta of prey animals; it is, nevertheless, an important factor in crocodilians' total nutrition.

Many, if not most, crocodilians exhibit a "feeding frenzy" when eating. If care is not exercised at this time, traumatic wounds may occur when one animal seizes another, mistaking it for a food item. Providing adequate space and avoiding overcrowded conditions will reduce the incidence of feeding-related trauma.

Table 4
Food Preferences for Selected Crocodilians

Crocodilian Species	M	Fish	Birds	Carrion	Frogs
American alligator *Alligator mississippiensis*	X	X	X	X	X
Chinese alligator *Alligator sinensis*	X	X	X	X	X
Caiman *Caiman sp., Melanosuchus niger*	X	X	X	X	X
Crocodile *Crocodylus sp.*	X	X	X	X	X
Gavial (gharial) *Gavialis gangeticus, Tomistoma schlegeli*	O	X	O	O	O

X = usual food items; O = occasionally eaten; M = miscellaneous meats.

Table 5
Food Preferences of the Tuatara

	Insects	Immature mice	Fowl chicks
Tuatara	crickets	X	O
Sphenodon punctatus	silk moth larvae		quail

Tuatara

The tuatara, rarely exhibited and then only in strictly authorized institutional collections, feeds on cultured crickets and live mouse pups which it swallows whole after brief mastication. Wild-caught crickets are not fed to these highly prized archaic reptiles because of the possibility of transmitting parasitic diseases that wild crickets may contain. Now that mulberry silk moth larvae have been shown to possess superior nutritive qualities to many other insects, they may be a useful substitute for crickets.

WATER REQUIREMENTS, DEHYDRATION, AND GOUT

The necessity to drink additional water depends largely upon the moisture content in the dietary items, "insensitive" water loss to the environment through the respiratory system, and the nature of an animal's urinary and fecal wastes. For example, aquatic turtles, all crocodilians, and some lizards excrete a fluid urine and soft, moisture-laden stools. Many terrestrial lizards and chelonians and most snakes excrete much drier feces and chalky urinary wastes from which much moisture has been extracted. Many desert dwelling reptiles are capable of recycling considerable amounts of water from the urinary wastes while they are stored in the bladder. Other reptiles do not possess these urinary reservoirs but, nevertheless, produce relatively dry renal wastes. These characteristics, together with the complementary action of salt-excreting glands (see below) in conserving water, make it possible for many reptiles to survive in some extremely harsh and arid habitats. A striking example of this parsimony was described in the Palestine saw-scaled viper, *Echis colorata*, by Goode (1983); specimens were raised in an arid environment without water for as long as 665 days. All snakes in the study group were fed mice every 7–21 days and remained completely healthy during the period of observation.

Much of a reptile's body water economy is derived from the moisture contained in its food, but most species will drink an additional volume of water *if* it is presented in an acceptable form and manner. Many reptiles will refuse to drink voluntarily from containers of standing water. It is imperative to know which reptiles will accept moisture only as dew droplets. Some will drink water only by lapping dew-like drops from misted foliage; others will readily drink from vessels of fresh water. Frequently, captive reptiles will soak for prolonged periods in their water containers. It has been reported that the Australian moloch lizard, *Moloch horridus*, is able to absorb water, in the form of rain, from its skin. More recent investigation has revealed that water is directed via tiny interscalar channels in the integument to the mouth, where it is swallowed. A similar mechanism of interscalar channeling of moisture collected by the spiny skin and cephalic horns is believed to function in horned lizards, *Phyrnosoma* sp.

Some reptiles possess highly specialized extrarenal sodium, potassium and chloride-secreting "*salt glands*." These glandular structures produce a concentrated or hypertonic secretion, thus aiding in the conservation of precious water stores by removing excess salts without the loss of significant amounts of aqueous solvent.

A relative or absolute deficiency of *metabolically available* water often leads to dehydration and the accumulation of insoluble sodium, potassium, and ammonium urate salts within the kidneys and other tissue sites. Replacement fluid should be administered at a dosage of 15–25 ml/kg every 24 hours.

In the case of terrestrial chelonians and some species of lizards, the concentrations of urates may promote the formation of urinary bladder stones which occasionally reach enormous size.

Because of the intimate relationship between water and the excretion of nitrogenous urinary waste products, it is important to understand the course of nitrogen-rich food sources after they are eaten. In most reptiles, the end products of such nitrogenous feeds are salts of uric acid. These urate salts are the biochemical residues resulting from the reaction of uric acid with sodium, potassium, and ammonium ions in the body fluids.

For some snakes and lizards, the plasma uric acid rises dramatically a day or two after eating a meal. It then quickly returns to a prefeeding level. In other reptiles, particularly those whose proteinaceous nitrogen is derived from plant sources, uric acid reaches the peak level and returns to the prefeeding state over a longer time. A reptile which is already moderately dehydrated is thus placed at a greater risk of uric acid accumulation than one which is properly hydrated.

The clearance of uric acid or urates by the kidneys can also be substantially diminished by any of several nephrotoxic drugs, particularly some antibiotics and poisonous plants. Because of the relatively primitive blood circulation to and from the kidneys of reptiles, potentially toxic agents should not be injected in the rear half of these animals.

As long as the blood circulation to the kidneys is adequate to maintain the clearance of urate salts, accumulation is unlikely to occur; however, because of the insolubility of these salts, many reptiles are placed at substantial risk of inflammation of joints and/or internal organs due to articular/periarticular or visceral gout after experiencing water deprivation. Thus, it is absolutely essential that fresh water is always readily available to all captive reptiles.

Salt Metabolism and Requirements

As noted previously, many reptiles possess extra-renal salt glands that secrete highly concentrated solutions that contain excess sodium, potassium, and chloride ions. Because these secretions contain greater concentrations of these salts than are present in the blood plasma, little water is lost when these electrolytes are eliminated. Most people who keep iguanas have observed that these lizards sneeze frequently; when the fluid expelled from the nostrils dries on clear glass surfaces, it can be seen to contain white crystals. These crystals contain sodium and potassium chloride and lesser amounts of other metabolites.

This mechanism has a great survival benefit in animals living in habitats where fresh drinking water is scarce; thus, this form of electrolyte excretion is commonly employed in desert-dwelling species. These glands also permit some marine animals to drink salt water because they are able to excrete excessive electrolytes without losing precious water.

Most diets contain adequate levels of sodium, but if there is doubt, some supplementation with iodized table salt can be furnished. Only very small amounts of sodium chloride are required because most rations fed to captive reptiles (even those whose diets consist of fruits and vegetables) contain traces of sodium and chloride. Animals who eat flesh obtain adequate amounts of sodium, chloride, and potassium in the diet and do not benefit from supplementation.

Frequency of Feeding

Some reptiles, especially snakes, some of the heavy bodied lizards, most crocodilians, and many chelonians tend to overeat and become morbidly obese in captivity. Because they no longer must forage for their livings and are fed too frequently or too much, they often accumulate enormous volumes of fat within their body cavity, in subcutaneous tissues, and even in some visceral organs such as the liver. The more active snakes such as racers, coachwhip snakes, garter snakes, vine snakes, mambas, and many cobras, etc., usually avoid obesity if they are fed once or twice weekly. A similar situation exists with many lizards. Agamas, anoles, basilisks, many geckos, swifts, iguanas, water dragons, whip-tail lizards, race-runners and some old world chameleons usually are active enough in captivity to avoid obesity and can be fed three to four times weekly.

Most aquatic turtles (with the possible exception of the more sedentary adult common and alligator snapping turtles) forage actively and usually do not become obese. These creatures can be fed almost daily. Most tortoises, if allowed to browse or graze daily on a variety of vegetation, flourish when fed daily. Of course, the nutritional value of a particular foodstuff plays a major role in health; poor quality feeds must be consumed in larger volumes to provide the necessary calories and digestible nutrients essential for growth and maintenance of the animal's tissues. Highly concentrated rations should be fed in smaller volumes. Desert tortoises and box turtles should have access to fresh fruits and vegetables to supplement their grass and leafy forage, which may be deficient in calories. There are exceptions but, in general, the more active animals may be fed almost daily. In contrast, the sluggish, mature, heavy-bodied pythons, many boas and anacondas, and many vipers and pit vipers should be fed only once every six-to-ten weeks. Some herpetoculturists feed their mature pythons, boas, and anacondas three or four times yearly. Additional exercise can be provided by giving the animals a tank in which to swim.

The larger tegu lizards, many monitors, Gila monsters, and Mexican beaded lizards, some giant skinks and similar heavy bodied lizards should be fed only once or twice weekly, depending upon the age of the animal and the nutritional quality and caloric content of the ration. Rapidly growing animals should be fed more often and greater amounts than mature animals of the same species.

Most mature crocodilians thrive when fed once or twice weekly. The gharials, feeding upon a diet mostly of fish (which has a high water content), must be fed more often.

2

Nutrition-Related Illness and Its Treatment

INTRODUCTION

A large percentage of the clinical conditions for which captive reptiles are presented to veterinarians are directly or indirectly related to the diet. Often, the herpetoculturist is unaware of the natural diet that is preferred by a particular reptile; other times, although the natural diet is known, it is unobtainable year-around and substitutes are attempted. Other times, the hobbyist is given incorrect information which results in macro- and micronutritional deficiencies or excesses which result in serious multi-organ dysfunction.

As mentioned in the introduction, this manual is not intended as a do-it-yourself veterinary textbook. Rather, the more commonly encountered nutrition-related conditions are discussed and their gross clinical characteristics are described so that the non-veterinary hobbyist can more easily recognize when a particular animal is abnormal. At that point, the author urges that professional veterinary medical assistance should be sought before the condition worsens.

In this chapter, vitamin and mineral deficiencies and excesses are described and methods for their prevention and treatment are suggested.

There are few commercial rations that have been formulated specifically for a particular group of reptiles. For instance, scientifically designed foods are available for horses, cattle, sheep and goats, dogs, cats, swine, primates, domestic fowl, game and exotic birds, and several types of fish. Each of these products was specifically created to meet the nutritional requirements of a particular animal. Because dietary requirements vary between species, many of these products are incompatible for non-target species and, in fact, may induce dietary imbalances if fed to reptiles.

ANOREXIA, INANITION, AND STARVATION

There are many reasons why any animal exhibits anorexia, the proper term to describe a lack of appetite or failure to feed: infectious or metabolic disease; parasitism; failure of the animal to adapt to the conditions of captivity; improper environmental temperature; too frequent handling and, of course, the wrong diet being offered.

Although the most common reason for anorexia is an inadequate or incorrect diet, environmental temperature also plays a major role. An environmental temperature which is too low to allow for normal activity of digestive enzymes will prevent even the well adapted reptile from displaying an appetite or from digesting or assimilating its food.

Most reptiles exhibit active feeding when the environmental temperatures are approximately 25–33 C (77–91°F). Of course, exceptions exist within the class, particularly with respect to the old world chameleons and the Tuatara, both of which prefer lower ambient temperature ranges. Similarly, other reptiles tolerate—and prefer—far warmer temperatures. Most snakes will refuse to feed just prior to shedding their senescent skin.

Approximately two decades ago, it was found that the antiprotozoan chemotherapeutic drug metronidazole (Flagyl-Searle), at a dosage of 12.5 to 25.0 mg/kg administered as a single oral dose, often will induce a previously anorectic snake to feed. (This is one-tenth of the recommended therapeutic dose for effective antiprotozoal treatment). The mechanism for this response is unclear, but the effect has been frequently demonstrated by many workers and does not appear to be directly related to the presence of intercurrent protozoan infections. The metronidazole is delivered directly into the stomach as a suspension with a tube catheter. This is a task that is best performed by a veterinarian possessing experience with reptiles.

Often a brief exposure to natural *unfiltered* sunlight (not through a window pane) outdoors together with a "shower" will induce many previously anorectic reptile to feed voluntarily.

A word of caution: Often, reptiles which have never exhibited overt aggression toward their keeper(s) will display marked changes in behavior after even brief exposure to natural sunlight outdoors. This aggressiveness usually subsides as soon as the reptile is taken back to its enclosure. Perhaps the most striking examples of this altered behavior have been observed in the Gila "monsters" and Mexican beaded lizards. Many previously docile animals actually bite those who have been unwary enough to casually pick them up after they have been permitted to sunbathe.

Reptiles, particularly snakes, are often intolerant of handling after capture and should be left undisturbed as soon as they are installed in suitable cages or after moving to new quarters. A hiding box or other refuge often proves useful in aiding the new captive to accept captivity and encouraging it to commence feeding.

In some instances, in order to provide necessary nutrients to an anorectic reptile, the surgical placement of an esophagostomy tube is warranted. Once in place, fluid nutrients can be delivered into the stomach with minimal stress to the patient.

Reptiles are functionally ectothermic and largely assume the temperature of their environment. Because of this, their basal metabolic rate usually is much lower than that of a similar size bird or mammal. Because of their lower rate of metabolism, reptiles can usually survive prolonged periods of fasting, but once they have crossed a poorly defined threshold, they decline in health very rapidly. Some species of snakes are well known for their habit of fasting to the point of starvation. At first, the affected animal lives off its stored fat, but soon this source of energy is exhausted and other tissues, especially muscle, are sacrificed to maintain life. The reptile slowly becomes gaunt and wasted and is more prone to infectious diseases and the stress of captivity.

As mentioned previously, this behavior is often related to a failure to accommodate to the conditions of captivity. One snake that displays prolonged fasting is the ball (or regal) python, *Python regius*.

Generally, if a reptile refuses to feed for a period of 6–8 weeks, its general health and the conditions under which it is being kept should be evaluated. If no physical evidence of disease is found, one or more hand-feedings should be given. The technique for this maneuver is provided below.

Forced Feeding

The snake is restrained so that its head and body are supported and its mouth is gently opened with a clean soft rubber or plastic kitchen spatula. A freshly killed mouse or other appropriate food item, lubricated with egg white, is placed into the mouth, using the rodent's pointed nose as a wedge. The food is gradually advanced into the back of the mouth and thence into the throat. A clean rubber eraser-tipped pencil can be used as a push rod to aid in passing the mouse down the esophagus until the meal is well down the snake's body. Once it has passed the head, it can be further advanced by gentle massaging from the outside, using the fingers of the operator.

MINERAL IMBALANCES, DEFICIENCIES, AND EXCESSES

Metabolic Bone Disease

Metabolic bone disease results from a calcium:phosphorus imbalance, calcium deficiency, phosphorus overload or, more rarely, an excess of metabolizable calcium (and vitamin D_3) in the diet. It is the most common mineral-related nutritional problem in captive reptiles.

In order to discuss metabolic bone disease, one must understand normal bone growth and maintenance, as well as the gross bone pathology. The common thread of vitamin D_3 must be woven into the fabric of this discussion, and it will be mentioned again later where it relates to other vitamins and minerals.

In nature, the calcium:phosphorus ratio of most, if not all, minerally adequate reptile diets usually falls within the broad range of between 1:1 to 2:1. One can appreciate how easy it is for an animal such as a snake or large monitor lizard to have ready access to diets with this level of calcium because these creatures consume their prey animal entirely; bones, muscle, fur or feathers, intestinal contents, visceral organs, etc. are swallowed; not a scrap is wasted because the prey is swallowed *in toto*.

Diets for captive reptiles which are rich in muscle meat and that lack bones (whether from mammals, birds or fish) may contain a calcium:phosphorus ratio of as much as 1:40. Herbivores fed diets consisting mostly of head lettuce, grapes, fruit, or mealworms receive similar imbalances. For this reason, many clinical cases of metabolic bone disease, formerly termed fibrous osteodystrophy, are seen in pet iguanas and aquatic turtles that have been fed insufficient calcium or, more importantly, far too much phosphorus in relation to available calcium. With such phosphorus overloading, it is difficult to balance the ration with compensatory calcium products. A deficiency of vitamin D_3 only exacerbates an already severe problem. Similarly, the feeding of newly born "pinky" mouse pups *that have not nursed recently* can result in gross imbalances in calcium and phosphorus intake. Data describing the mineral content of young and adult (30 gram) mice was published by Allen et al., (1982) and Fowler (1978), respectively. This information was reviewed further by Barten (1982) and is reproduced in Table 6.

As can be seen, new born mouse pups, because of their poorly mineralized skeletons, are deficient in available calcium; it is not until they have reached at least one week of age that their bodies exceed a calcium:phosphorus ratio greater than one. Barten (1988) noted that the feeding of new born mouse pups to diminutive rodent-eating snakes has proven to

Table 6
Mineral Content of Mice

Age of Mice	1–2 Days*	1 Day*	7 Days	Adult
Calcium (%)	1.6	1.72	1.43	0.84
Phosphorus (%)	1.8	1.66	1.29	0.61
Ca:P Ratio	0.9:1	1.0:1	1.1:1	1.4:1

*The first and second columns represent different groups of mice in Allen et al.'s study.

be effective even in the absence of vitamin D_3 supplementation and/or ultraviolet light irradiation. Barten suggested that, although some reptiles might experience metabolic bone diseases if they are fed only newborn mouse pups, calcium deficiency usually is avoided if the pups are fed before being offered to reptiles, probably because the calcium-rich mouse milk that had been ingested by the mouse pups is also available to the reptile.

The species, age, sex and size of some reptiles predispose them to the induction of inadequate skeletal mineralization. In a severely affected reptile, the bones become so depleted of mineralized matrix that skeletal deformities result. As the animal tries to bear weight on its limbs, the bones are unable to resist the forces exerted upon them and collapse or fold like bent twigs or soda straws.

The mandibles of lizards and crocodilians frequently are distorted by the rear-ward tension exerted by the tongue musculature; the softened mandibles, unable to resist this retraction, tend to bow outward. Turtles and tortoises fed bone-weakening diets also display gross deformities of their softened carapaces and plastrons. These deformations result from the downward tension on the underside of the carapace in the region of the attachment of the brachial and pelvic girdles and their associated musculature. Typically, these deformities assume the shape of shallow concavities in the bone and shell overlying each of the four limb attachments.

Radiographs of the skeleton of a reptile suffering from metabolic bone disease disclose a marked thickening of the cortical bone, but at the same time, a concomitant loss of mineral density.

This widening is an attempt to compensate for the overall weakness of the bones' structure by adding quantities of poorly mineralized bone matrix and connective tissue to the cortex. These grossly altered bones are prone to spontaneous fractures and distortion.

The cross-sectional diameter of the marrow cavities in the long bones appear little changed from bones of equal size from a normal animal.

Treatment

1. Oral or injectable calcium supplements such as calcium gluconate or calcium lactate should initiate the treatment for this condition. Calcium carbonate may be added to the ration. Ground hen's egg shells can be used to furnish calcium. A large spoon can serve as the mortar, and a smaller spoon is used as the pestle to grind the shells. Several available proprietary products contain only the calcium salt without accompanying phosphorus. A small amount of Vitamin D_3 usually is added to these products and enhances calcium absorption from the intestine. When treating clinical cases of metabolic bone disease with insufficiently mineralized bones, oral vitamin D_3 in appropriate dosage should be instituted at once. Although it may contain some phosphorus, Pet Cal (Beecham-Massengill) is, nevertheless, very useful in the management of the form of metabolic bone disease known as fibrous osteodystrophy (or secondary nutritional hyperparathyroidism) and is readily accepted by many reptiles, particularly turtles, crocodilians, and lizards.
2. A change of diet from one which originally induced the disorder to one which contains the correct amount of available calcium plus sufficient, but not excessive, phosphorus is recommended. Some foodstuffs including boned meats, iceberg lettuce, grapes, and mealworms actually exacerbate a calcium deficiency by con-

taining an excessive amount of phosphorus with respect to calcium. Other foods contain substances which bind to calcium and, thus, make it unavailable for normal mineral metabolism. Among these are spinach, which contains oxalic acid, and oats, which contain phytic acid. Both of these chemicals can actually make an already calcium deficient animal develop even more serious bone loss and skeletal deformity.
3. Exposure to a source of either natural *unfiltered* sunlight or artificial ultraviolet light of appropriate wave-length is necessary for the internal synthesis of vitamin D_3.

Rickets

Rickets is another form of metabolic bone disease characterized by softening of the bones. In this disease, the available dietary calcium may be sufficient, but there is a dietary or environmental deficiency of vitamin D_3. Either the preformed molecule is not present in the food or the animal is not exposed to suitable ultraviolet irradiation. The end result is identical in that the calcium is not absorbed from the intestine.

Rather than forming bone matrix by the orderly mineralization of cartilage "templates" at the ends of growing bones, the bones of animals suffering from rickets tend to undergo a variable degree of cartilage overgrowth called hypertrophy (abnormally large cell *size*) and hyperplasia (abnormally large *numbers* of cells). Such bones remain very poorly mineralized. One characteristic of this disorder is the formation of swellings termed "rachitic rosettes" at the junctions where the ribs join the breastbone. In addition, the ends of the long bones may be greatly thickened, or buttressed. The articular ends next to adjacent joints become broadly flared.

Radiographs of affected bones reveal little or no evidence of mineralization. The outer or cortical bone also may exhibit marked thinning. Folding and compression fractures also are frequently encountered, particularly in weight-bearing limb bones and vertebrae, respectively.

Renal-Associated Fibrous Osteodystrophy ("Renal Rickets")

The radiographic findings of this kidney-related skeletal disorder may be strikingly similar to those of secondary nutritional hyperparathyroidism, but the underlying cause is a loss of calcium by the diseased kidney with a corresponding retention of phosphorus. As the plasma phosphorus rises, the parathyroid glands, tiny paired structures in the neck or just in front of the heart, are stimulated to secrete ever-increasing amounts of parathormone, which causes the calcium stored in mineralized bone deposits to be leached out. The result is a markedly weakened skeleton. Renal-associated metabolic bone disease can be differentiated from dietary-induced secondary nutritional hyperparathyroidism by appropriate kidney function testing and determination of plasma creatinine.

Primary Hyperparathyroidism

To the date of this writing, only five confirmed cases of primary hyperparathyroidism have been documented in reptiles (Frye, 1981, Frye, 1986; Frye and Carney, 1975; Ippen, 1965). Four were in tortoises and one was in an iguana; all were directly related to hormonally functional benign parathyroid tumors.

The clinical signs are similar to those of both secondary nutritional and renal-associated hyperparathyroidism, but kidney function tests and/or investigation into the diets fed to these animals help rule out these causes.

A discussion of calcium metabolism in lizards would be incomplete without at least a brief mention of the intra- and extracranial endolymphatic duct and sac system that has evolved in some geckos, some (anoline) iguanids, and at least one small chameleon lizard, *Brooksia*. (See Bauer, 1989; Bustard, 1967, 1968; Dacke, 1979; Fitch, 1970, 1980; Gardner, 1985; Ineich and Gardner, 1989; Kästle; Kluge, 1987; Moody, 1983; Packard, Packard, and Boardman, 1982; Packard, Packard and Gutzke, 1984; Packard et al., 1985; Ruth, 1918; Simkiss, 1967; and Werner, 1972). This fascinating structural system originates from the saccules of the inner ear and terminates in expanded sacs which lie between the mininges (Whiteside, 1922). These sacs fill with calcium carbonate and can be mobilized during the formation of yolk and egg shell formation. As Bauer noted, the yolk of snake and lizard eggs contains more calcium than that of bird eggs (Packard et al., 1984) and ultimately provides the majority of the calcium requirements of the embryo (Packard et al., 1984, 1985) and probably also during early post-embryonic life. The egg shells of many gekkonid lizards are more heavily calcified than those of many other lizards, and thus, appear to require a greater degree of calcium mobilization during their formation.

Miscellaneous Calcium-Related Disorders

Excessive Calcium in the Diet

Only when accompanied by an excess of vitamin D_3 is an over abundance of calcium a problem. Some amateur herpetoculturists have accidently overdosed their reptiles by feeding vitamin-mineral supplements at the same time that potent sources of vitamin D_3 were being added to the diet. To further worsen the situation, the animals were exposed to both unfiltered natural sunlight *and* artifical ultraviolet cage illumination. The result of this situation is a vitamin overdosage. This condition causes greater than normal amounts of calcium to be absorbed from the intestine. Then the abnormal plasma calcium induces the deposition of calcium-rich mineral salts in soft tissues that are not normally mineralized. The aorta, heart muscle, pulmonary airways, gastrointestinal, and genitourinary tissues are particularly sensitive to this visceral mineralization. Fortunately, unless excessive vitamin D is present, inordinate dietary calcium passes through the intestine with little absorption and, therefore, there is little or no increase in plasma calcium.

The Effects of Feeding Excessive Amounts of Dog, Cat, or Monkey Chow

The feeding of *small* amounts of commercial dog, cat or monkey chow to some captive reptiles has been recommended by some authorities, including this author. Because these commercial pet diets are balanced for dogs, cats or monkeys, the amounts of available vitamin D is usually greatly in excess of that required—and tolerated—by most reptiles.

When these rations are fed in moderate to large amounts to some reptiles, they induce severe, often fatal, soft tissue mineralization involving many organs and tissues. Organs containing smooth muscle are particularly sensitive to these alterations. After being fed a diet with abnormally high levels of vitamin D, some reptiles will yield plasma calcium levels that may exceed 40 mg/dl; this is approximately three times the level recorded for many normal reptiles. With prolonged high plasma calcium levels, soft tissues undergo severe and often massive mineralization. It is interesting that there is a marked sex difference in the clinical expression of this diet-related disorder. Matched pairs of iguanas, for example, may be fed identical diets, rich in vitamin D_3 and calcium; mature females, because of their egg shell production, may be spared soft tissue mineralization whereas their male partners may be very severely affected.

Because of their convenience, availability and palatabilty, great care must be taken to refrain from feeding these commercial feeds in excess.

Treatment for Clinical Hypercalcemia

Hypercalcemia can be effectively treated with the hormone, calcitonin. To the date of this writing, the dosage of this bioactive agent has been scaled from that reported in the literature for dogs and cats that have been intoxicated with rodenticides containing vitamin D and its analogs.

The product that has exhibited efficacy in the treatment of hypercalcemia is Calcimar synthetic salmon calcitonin (Roerer Pharmaceutical Co., Inc.). Calcimar is supplied in a strength of 200 international units/ml. For use in most reptile patients, this product is diluted with sterile physiologic fluid (saline, Ringer's, etc.) to a final concentration of 1.0 international unit/ml. The subcutaneous dosage that has proven to be effective is 1.5 international units/kg (Dr. B.V. Centofani, personal communication). Adjunctive treatment to achieve enhanced urination is essential. Sterile physiologic fluids are administered intravenously, subcutaneously, or into the body cavity daily at a dosage of 15 ml/kg/24 hours. Thus, for an adult common iguana weighing approximately 2 kg, the dosage is 3 units administered three times daily plus 30 ml of fluid once daily until the plasma calcium returns to approximately 13 mg/dl.

Characterstically, the reduction of plasma calcium is abrupt and sustained provided that the calcitonin and diuresis are continued, but a rebound almost to pretreatment levels may occur if the therapy is discontinued prematurely. Of course, the animal must no longer be fed a diet high in vitamin D such as dog, cat, or monkey chow.

Trace Minerals

Iron, manganese, magnesium, cobalt, etc. are all, more than likely, required by reptiles, but have not been reported as deficiency states.

A well balanced vitamin-mineral supplement should be fed routinely and will, undoubtedly, prevent trace mineral deficiencies. The same situation appears to be true of selenium, which should be studied to elucidate its role in the micronutrition of reptiles.

Trace mineral deficiencies, when they are diagnosed in reptiles, occur most often in captive iguanas fed vegetarian diets limited to only a few fruit and vegetable species.

VITAMINS AND VITAMIN DEFICIENCIES

Vitamin A Deficiency

Vitamin A deficiency in reptiles is most frequently observed in captive semi-aquatic turtles, particularly juveniles. The yolk remaining at the time of hatching usually furnishes the vitamin A required by the rapidly growing youngsters for approximately six months even if they are fed a vitamin A deficient diet. Unless a source of vitamin A is ingested either as preformed vitamin (found in animal tissues) or sources of the precursor beta carotene (found in green, yellow or orange vegetables), the neonatal yolk-derived vitamin A is slowly expended from the liver where it was stored. Once a threshold is crossed, the clinical signs of vitamin A deficiency can be expressed with alarming speed and increasing severity.

The mucus-secreting glandular structures associated with the eyes, pharynx, upper airway and lower respiratory passages rapidly undergo and display tissue alterations termed squamous metaplasia; mucus production is diminished, and swelling and overgrowth of the horn-like keratin is manifested. The eyelids, moist conjunctival membranes lining the eyes and other ocular structures, the endocrine glands, and the genitourinary tract also are involved early in the course of this disorder. The eyelids become swollen and may become inflamed. The tissues lining the mouth become overgrown and tend to accumulate as chalky, scale-like deposits. These deposits further impede the function of the jaws and make the grasping of food and swallowing difficult. Such mucous tissues are prone to infection and injury.

High protein diets deplete neonatal stores of yolk-derived vitamin A more rapidly than low protein rations. Affected turtles usually have swollen eyes and, often, signs of respiratory distress. The severity of these signs generally correlates with the degree of vitamin A deficiency and the length of time that the animal has been fed a vitamin A deficient diet. As the condition becomes chronic, the widespread effects become more severe and successful treatment becomes difficult to achieve.

Treatment consists of oral or injectable vitamin A. (See description of treatment-associated excess vitamin A, below.) Aquasol A (U.S. Vitamin and Pharmaceutical) is water miscible and a dosage of 10–10,000 Int. Units is indicated depending upon the size of the patient and the severity of the lesions. Oral administration of a product such as ABDEC Drops (Parke-Davis) may be given on a regular basis and/or for correction of the diet responsible for the deficiency. Obviously, the diet must be corrected to include an adequate level of vitamin A. Here is an example where a recommendation can be made for the feeding of commercial dog or cat food (particularly the semi-moist cat foods like Tender Vittles (Ralston Purina). These foods must be fed in modest quantities only. Another excellent diet which contains adequate levels of preformed vitamins and minerals is Trout Chow (Ralston Purina). Both of these products are palatable to small turtles, which readily eat them. The diet for these small aquatic turtles may be further supplemented by feeding earthworms and small live or freshly killed fish.

Vitamin A Overdosage

It is a common practice in clinical reptile medicine to employ injectable vitamin A, either as a sole bioactive agent or as a multiple oil-soluble vitamin preparation containing vitamins A, D, and E (as alpha tocopherol or as mixed tocopherols). Often the actual need for the patient to receive these supplemental vitamins is not adequately evaluated and their administration becomes "routine." Often the patient is an herbivorous terrestrial tortoise whose normal vegetarian diet contains abundant natural carotenoids. Sometimes the clinical sign that is seized as a justification for using these drugs is a nasal discharge. Occasionally these vitamins are injected with the expectation of improving a flagging appetite. This last usage should be condemned because 1) there is no evidence for such an effect and, 2) it often induces more disease than it cures!

The clinical signs of vitamin A excess in terrestrial chelonians are the rapid development of dry or flaky skin called subacute xeroderma followed in a few days by redness (erythema), inflammation, and severe loss of skin. Once the skin covering has been lost, the underlying muscle tissues become exposed and are easily infected. These lesions typically appear on the limbs and skin of the neck and tail, but they may occur on any skin-covered surface. Usually by the 14th post-injection day, the epidermis has formed large fluid-filled blisters, called bullae, which lift away from the underlying dermis and occasionally from skeletal muscle, variable-size expanses of denuded moist tissue.

Interestingly, these alarming clinical signs appear to be identical to those reported and illustrated in the lay literature as being induced by the antibiotics carbenicilin and gentamicin sulfate. It is not clear whether all of the cases that have been described also had received vitamin A injections as part of their treatment.

Eventually the dermis and muscular structures are exposed. Most of the tortoises affected with this disorder survive, and their damaged skin will regenerate. Others, however, resemble patients suffering from third degree thermal burns and die as a result of the effects of having such large areas of their skin destroyed; tissue fluid loss and infection soon exact their toll.

Prudence suggests that the use of injectable vitamin biologicals in reptiles should be avoided *unless* the dietary history and clinical signs displayed by a particular animal clearly signal that a deficiency exists. This appears to be particularly germane in the case of herbivorous tortoises which ingest (and can store) large quantities of beta carotene, the precursor of vitamin A.

If clinical judgement suggests that there will be a benefit to the use of supplemental vitamin A, *oral* dosage with natural sources of carotene and/or preformed vitamin A appears to be the safest means for administration. Freshly grated carrots, yellow squash, green leafy vegetables, and alfalfa are readily available and highly palatable.

Vitamin B_1 (Thiamin) Deficiency

Most often, a deficiency of thiamin is a result not of insufficient preformed or internally synthesized thiamin but, rather, the action of the lytic enzyme, *thiaminase*, which is present in the flesh of dead fish, some clams, and certain plants consumed by captive reptiles. Under most conditions, the intestinal microflora synthesize water-soluble B-complex vitamins which are then absorbed from the stool contents as they traverse the terminal digestive tract. The feeding of thawed, frozen fish of several species, some clams, and several varieties of vegetation may result in thiamin deficiency, followed by signs of central nervous system and/or cardiac abnormalities in a variable length of time, depending upon the amount of thiaminase contained in the food items and available thiamin in the diet and host tissues. The longer the deficiency state is allowed to exist, the more dramatic will be the clinical manifestations. The induced nervous disorders usually are seen as postural and behavioral alterations, anorexia, muscle tremors, and, eventually, death. Blindness may be seen, but this is an infrequent sign.

Treatment consists of the administration of injectable thiamin hydrochloride and the correction of the dietary source of thiaminase that produced the deficiency. Rather than feeding previously frozen fish, one can substitute live goldfish or similar species. Dead mice and rats may be disguised as fish by merely rubbing some fish slime over their fur before offering them to fish-eating snakes. Thawed, frozen fish may be altered by inserting one or more thiamin HCl tablets into their body cavities *just prior* to feeding them to the snakes or other fish-eating reptiles. The dosage is not critical and approximates 25 mg/kg during the initial treatment phase which lasts 3–7 days. Eventually, correction of the diet should be sufficient for maintenance without supplementation.

In the case of herbivorous species which consume plants containing thiaminase, other plant sources must be substituted. Prolonged antibiotic therapy may also induce thiamin deficiency by reducing or eliminating the intestinal microflora which synthesize thiamin. This should be borne in mind when caring for reptiles on chronic antibiotic therapy. The reduction or loss of microflora often may be remedied by feeding fresh yoghurt or *Lactobacillus acidophilus, L. planarum,* and *Streptococcus faecium* culture products. Also force feeding small volumes of a slurry composed of finely ground alfalfa pellets has shown merit. These reinoculation treatments are particularly useful in treating chelonians and herbivorous lizards which are usually the types of reptiles affected with this nutritional disorder.

Biotin Deficiency

Biotin deficiency is similar to thiamin deficiency because it is an induced condition caused by abnormal diets. In the case of biotin, the usual cause can be traced to the feeding of raw hen's eggs to the exclusion of almost all else. Raw avian egg albumen contains a substance, avidin, which possesses an antibiotin biologic action.

In nature, egg-eating species such as the Gila monster, *Heloderma suspectum*, the Mexican beaded lizard, *Heloderma horridum*, the Tegu, *Tupinambis teguixin* (or *T. nigropunctatus*), several species of monitor lizards, *Varanus* sp., and the egg-eating snake, *Dasypeltis scabra*, usually encounter and consume *fertilized* eggs. This is an absolutely vital distinction because the embryonic tissues present in the developing eggs contain biotin. Under most captive management situations, *unfertile* eggs are fed which, by mere definition, lack embryos and contain little biotin. Furthermore, much of the avidin present in the embyronated egg is expended as development progresses. Lastly, many, if not most, of the oviphagous species of reptiles readily attack and devour small mammals and birds as well as their fertilized eggs; these dietary items are a rich source of biotin.

Ascorbic Acid (Vitamin C) Deficiency

Available evidence to date has shown that most reptiles are capable of synthesizing vitamin C in their kidneys and/or intestines (Chatterjee, 1967; Chatterjee, 1970, 1973; Vosburgh et al., 1982).

It has been suggested that intercurrent renal or intestinal diseases diminish ascorbic acid synthesis and that under some conditions of captivity and/or chemotherapy, a state of *induced* vitamin C deficiency may occur. Some investigators have pointed to the fact that much preformed vitamin C present in the intestinal contents of small animals fed to reptiles may have already passed in the feces prior to consumption by the reptiles. Simply feeding the prey before offering them to the reptiles is an expedient means for preventing vitamin C deficiency. Alternatively, ascorbic acid may be inserted into freshly killed (or thawed frozen) prey animals just before offering them as food.

The clinical manifestations of vitamin C deficiency are usually referable to reduced collagen synthesis or tissue (particularly skin and vascular) strength. Sudden, acute skin rupture and/or spontaneous gingival bleeding are two common signs.

Typically, the therapeutic response of reptiles to remedial therapy is rapid.

Vitamin D Deficiency (Particularly D_3)

The intimately related roles of calcium, phosphorus and vitamin D_3 were discussed earlier in the section dealing with metabolic bone disease. The absolute or relative deficiency of vitamin D_3 results in diminished calcium absorption from the intestinal contents. With decreased plasma calcium levels, mineralized bone matrix formation is retarded or arrested entirely, resulting in rickets. It must be remembered that for calcium to be absorbed from the intestine, vitamin D_3 must be present, either from the internal synthesis of the vitamin through exposure of the animals to ultraviolet irradiation, or via ingestion of the preformed active vitamin. Because of the potential for excess vitamin D and its toxic effects, including massive and widely disseminated soft tissue mineralization, vitamin D overdosage must be avoided.

A number of excellent proprietary sources of both calcium salts and vitamin D_3 are available. One which has been used successfully for many years is Pet Cal (Beecham-Massengill). It is palatable to many herbivorous reptiles and is available in a relatively hard-milled tablet form which can be broken into suitable portions. Because it dissolves slowly in water, many aquatic turtles will readily nibble at the softened fragments. As Barten (1988) suggested, birds also require vitamin D_3; therefore, vitamin supplements compounded for avian species are appropriate for use in reptiles.

Hen's egg shells, coarsely ground cuttlebone, calcium carbonate, and similar natural calcium-rich materials are suitable alternatives.

Vitamin-E Deficiency

Like so many of the other deficiencies, hypovitaminosis-E usually, if not always, results from the feeding of an unnatural diet lacking in variety and rich in oil-laden or fatty fish or other sources of either highly saturated rancid or polyunsaturated edible oils. In one case diagnosed in a boa constrictor, the snake had been fed only obese laboratory rats. Interestingly, these rats had been fed an unnatural diet of only hulled sunflower seeds and water and were enormously fat; the body fat of these rodents had a near-liquid consistency.

Vitamin E is required for its antioxidant properties. The relationship between vitamin E and selenium is still under investigation, but it is now recognized that a synergism exists between the two substances acting as bioactive antioxidants. Deficiencies of either may induce similar lesions, particularly in muscle tissue; under conditions of captivity, a combined deficiency of both may be encountered. In cattle and other ruminants a positive therapeutic effect is seen only when both vitamin E (mixed tocopherols) and selenium are administered (Whitehair, 1970).

A selenium-vitamin E deficiency related muscle disorder, often called "white muscle disease," has been diagnosed in lizards (Frye, 1981; Rost and Young, 1984; Farnsworth et al., 1986), snakes, and aquatic chelonians (Frye, 1981). Skeletal and heart muscles are most often involved, but other tissues containing smooth muscle may be affected also.

In an interesting case, an iguana that exhibited clinical signs of hypocalcemia responded well to vitamin E and selenium therapy (Farnsworth et al., 1986). Prior to treatment, this lizard was affected with periodic spasmodic muscular tremors which often have signalled a calcium deficiency, but its pretreatment serum calcium was 13.2 mg/dl (normal is 9.6–14.3 mg/dl). After treatment with a vitamin E-sodium selenate preparation, Seletoc (Burns-Biotec), the iguana's muscular stiffness and tremors diminished gradually and it was normal after two weeks of treatment.

A thorough history and evaluation of the captive care and the diet fed may reveal a dietary source for the deficiency. Where selenium is deficient in the soil in which forages and seeds are grown, herbivorous animals are at risk of developing a muscular dystrophy characterized by muscle fiber swelling, loss of nuclei, and eventual necrosis and replacement of affected muscle tissue by fibrous connective tissue. Typically, most forages and seeds contain selenium salts of approximately 0.2–0.4 parts per million on a dry weight basis; deficient diets may contain less than 0.1 PPM (Rost and Young, 1984; Whitehair, 1970).

The diagnosis of muscular dystrophy is confirmed by biopsy.

Clinically, the affected animal suffering from chronic saturated lipid-induced vitamin E deficiency tends to have grossly abnormal fatty tissues, often appearing on palpation to be unusually firm and unyielding. The skin overlying such deposits of altered fat may have a yellow or orange hue, and on biopsy or necropsy, possess a rancid or distinctly fishy odor.

Treatment consists of the administration of injectable and/or oral alpha- or mixed tocopherols. The individual dosages have not, as yet, been determined for reptiles, but these agents (except for selenium) are, except in gross overdosage, minimally toxic; selenium has been shown to be toxic in amounts that exceed therapeutic quantities only slightly. Farnsworth et al. administered 0.025–0.50mg sodium selenate and 50–100mg vitamin E to a 1.8kg common green iguana and achieved an apparently excellent therapeutic result. This dosage was arrived at empirically (Farnsworth et al., 1986).

Vitamin K Deficiency

Except in accidental or malicious poisoning with any of the natural or synthetic coumadin derivatives (Warfarin sodium), vitamin K deficiency is almost always the terminal result of long-term antibiotic therapy which has caused a loss of the normal intestinal microflora which synthesize vitamin K. It also may occur spontaneously without clear causal factors. An example of the latter incidence was seen in some long-term captive crocodilians that, after many years on a mixed diet of fish, meat and fowl (which would appear to be an adequate fare for these opportunistic feeders) exhibited prolonged bleeding of the gums after shedding their teeth.

Treatment for hypovitaminosis-K consists of administration of oral and/or injectable vitamin K homologs. A varied diet also appears to help reduce the incidence of spontaneous bleeding. In nature, the crocodilians obtain significant amounts of vegetable matter in the form of accidentally ingested algae, weeds, grasses, and stomach and intestinal contents of their prey species.

Iodine Deficiency and Thyroid Dysfunction

Clinical iodine deficiency is most often encountered in terrestrial tortoises, particularly those giant taxa originating in the Galapagos and Aldabra Islands. This incidence suggests that these chelonians may have an evolution-related high metabolic requirement for halogens, particularly iodine. Many volcanic islands are noted for their halogen-storing flora, and it is possible that over thousands of years, selective pressures favored plants that could absorb and store relatively large amounts of iodine, chlorine, bromine, fluorine, and their respective metabolically active salts. Thus, tortoises faced with a limited flora consisting of halogen-sequestering plants, may have evolved to exploit the resources available to them.

In these chelonians, the clinical manifestations are those of fibrous goiter and appear to be related to *both* insufficient dietary iodine, *per se*, and also to feeding goiter-causing plants in the captive diet. Examples of these plants are: cabbage (essentially all varieties), kale, broccoli, Brussels sprouts, cauliflower, etc. If these vegetables are to be fed as a significant portion of the ration, some form of additional iodine should be provided in the form of sodium or potassium iodide or iodinated casein. The iodide solutions tend to be bitter, and the food to which they are added may be rejected by some reptiles. Ground kelp tablets are a far more palatable and natural source of supplemental dietary iodine. These sources of iodine are applied as top dressings to day-old bread or fresh grass and should be administered two to three times weekly. The dose is not critical and the iodide solutions and kelp are minimally toxic, except in gross overdosage. An adult box turtle on a goiter-causing diet, for instance, would benefit from 1–2 mg/daily for two to three weeks and then about that amount weekly as a maintenance dose. Larger or smaller reptiles should have their dosage scaled up or down from this amount.

HYPOGLYCEMIA IN CAPTIVE CROCODILIANS

A number of investigators have reported stress-induced hypoglycemia in crocodilians. The clinical manifestations are tremors of the fine and large skeletal muscles, slow-to-absent righting reflex, torpor and, almost invariably, widely dilated, non-light responsive pupils (mydriasis).

Treatment consists of orally administered glucose at a rate of 3 grams/kg, and the removal of the stressful conditions under which the hypoglycemia was induced (Wallach, 1971).

CONSTIPATION, OBSTIPATION, AND GASTROINTESTINAL BLOCKAGES

The feces of captive reptiles vary widely in their consistency, content, and frequency. There are no specific guidelines with which to judge whether a particular animal is constipated. The digestive process depends substantially upon the food content, environmental temperature and, consequently, body temperature of the animal, and length of time since its last meal. An example would be feeding rodents of different ages to a boa constrictor. Fully mature rats, because of their denser fur and skin and larger, well mineralized skeletons and teeth, are digested more slowly than a meal of an equivalent weight of hairless baby rats (which would also represent a far greater surface area per unit volume upon which the digestive secretions must act). If water is not readily available to the reptile, the stool masses may become dry, thus impeding their passage through the intestines, particularly the colon.

Often, the captive reptile suffers from the inability to complete the passage of ingesta through the digestive tract. In captivity, exercise may be severely curtailed, diets may lack sufficient roughage, and some of the animals ingest large amounts of cage litter. It is not uncommon for some captive reptiles to become grossly obese. All of these factors favor intestinal blockages and constipation, and must be addressed.

Many herpetoculturists employ pelleted alfalfa rabbit or guinea pig chow as cage litter. When this material is purposefully or accidentally ingested, it provides essential nutrients and safe fiber. When swallowed, it soon softens and is digested or at least passed through the alimentary tract. The non-nutrient fiber contributes to the stool bulk and is passed in the feces without inducing blockages.

Most frequently, intestinal blockages affect terrestrial chelonians. When a tortoise has suddenly stopped eating, it is not unusual to find that no feces have been passed for several weeks. With the retention of fecal material in the large bowel, there is a progressive absorption of fluid from the stool boluses. Soon, these boluses become abnormally dry and hard; normal peristaltic waves can no longer move them. Radiography at this point may reveal a physiological blockage called an ileus, with gas-filled loops of intestine just in front of the obstruction.

Treatment ranges from bathing the animal in tepid water (which often is sufficient to induce defecation) to major surgery for physically removing masses of dried stool and foreign material from the intestine. Medical management of constipation often produces satisfactory resolution. Small amounts of petroleum jelly, mineral oil, magnesium oxide (milk of magnesia) suspension, dilute dioctyl sodium sulfosuccinate (DSS), phenolpthalein, and/or stool bulk augmenters such as hemicellulose, may be sufficient. Siblin (Parke-Davis) is an easily available medication. Several proprietary laxative products formulated for cats are readily available from veterinarians. A small volume of any of these products is placed into the mouth of the affected reptile. These solid laxatives are less likely to be inhaled than liquid mineral oil. When using DSS, it must be diluted with warm water to a concentration of no more than 1:20 (based on the commercially available products). Some brands of DSS, when given in concentrated form, have been shown to induce severe, usually fatal tissue damage in the esophagus, stomach, and intestine (Paul-Murphy, Mader, Kock, and Frye, 1987).

This dilute solution of DSS may be delivered through a stomach tube and/or gentle enema introduced into the copradeum. If the latter method is employed, great care must be exercised to avoid forcing the enema solution into either the terminal urinary or the genital tracts.

If large volumes of gas are seen on radiographs, a small amount of simethicone solution should be given orally. This substance will aid in breaking up intestinal gas bubbles. Several proprietary products contain simethicone, but I have found that one, Mylicon (Stuart), produces consistently safe and effective relief from functional ileus-induced blockages due to intestinal gas. See bloating (tympany) later in this chapter.

VOMITING

This medical problem is included because it is so intimately intertwined with nutrition in general. The diagnoses of and causes of vomition in reptiles are essentially the same as those observed in higher vertebrates, but also include causes that are seen exclusively in reptiles. Improper environmental temperature; infectious and/or metabolic disease; parasitism, particularly cryptosporidiosis in snakes; intoxications; putrefaction of ingested material; foreign bodies or ulcerative lesions within the gastrointestinal tract; pyo-granulomata and abscesses, or tumors either involving the walls or impinging upon the walls of the gastrointestinal organs; and mere gorging are but some of the myriad number of causes for a reptile to vomit. A proper diagnosis requires obtaining a thorough history and evaluation of captive husbandry practices. Diagnostic radiography or other imaging techniques, microscopic cytologic examination of gastric lavage (stomach wash) specimens, and fiberoptic endoscopy, with or without simultaneous gastric biopsy, are diagnostic options which may reveal the cause of chronic vomiting. Analysis of the feces may disclose severe gastrointestinal parasitism.

An adequately warm ambient temperature must be provided when feeding captive reptiles. The underlying reason for this is that it permits and even enhances the proper digestion and assimilation of ingested food. This process is linked to several temperature-dependent digestive pancreatic, gastric and intestinal enzymes. Low environmental temperature does not permit internal body temperature sufficient to support digestion. The food simply putrefies and is soon vomited.

Handling a snake or lizard soon after it has eaten often results in the food being regurgitated. This behavior may be a defensive behavior evolved to discourage predators.

In the chelonian, regurgitation is an unfavorable prognostic clinical sign and should be evaluated by a veterinarian who is familiar with reptiles. A thorough review of the animal's history and physical examination findings, radiography and, when appropriate, analysis of blood chemistry may be required in order to arrive at a correct diagnosis. In practice, the first two diagnostic methods will usually reveal the cause of chronic vomiting.

DIARRHEA

Florid diarrhea is an uncommon problem in captive reptiles. When loose stools are passed by an animal, they should be examined for the presence of protozoan, metazoan, bacterial, and/or mycotic organisms. If any of these tests reveal disease-causing organisms, appropriate medication should be administered together with supportive therapy. If none of these organisms is found to be responsible for producing the abnormally soft stools, appropriate dosages of Kaopectate (Upjohn), Kaopectate Concentrate (Upjohn), Pectolin (EVSCO), etc., can be administered to help relieve the condition related to the size and weight of the individual animal. Sometimes loose stools are the result of the reptile's diet. For example, cucumbers and melons fed to many tortoises often induce very soft feces. Merely changing the diet to a less hydrated food may resolve the diarrhea. Adding pelleted ground alfalfa to the diet often firms the stools dramatically.

With continued diarrhea, the patient may become dehydrated and electrolyte imbalances can develop. The dehydration and electrolyte imbalances should be managed by the administration of replacement fluid and electrolyte solutions such as Entrolyte (Beecham Massengill).

The causes and treatment for vomiting were discussed in the previous section. Because the physiological effects of diarrhea and vomiting can be serious, every effort should be made to diagnose and treat these disturbances early, specifically, and aggressively. Moreover, because some particularly virulent disease causing organisms can induce both of these disorders simultaneously, a definitive diagnosis is important to reduce the likelihood of horizontal transmission within an animal collection or colony.

BLOATING (TYMPANY)

The production of excessive intestinal gas is most often observed as a clinical problem in herbivorous or omnivorous lizards and tortoises that have ingested food items containing easily fermentable fruit sugars or other substances upon which gastrointestinal microflora subsist. Often, this condition follows a too rapid change of a diet high in fiber to one containing a higher percentage of metabolizable carbohydrates. Typically, the affected reptile presents with grossly observable swelling of its abdomen. In the case of tortoises, the limbs are extended and the soft tissues covering the limb pockets appear to be protruding from beneath the shell. This condition can prove fatal if the lungs are compressed by the expanded gastrointestinal organs. Open-mouth breathing and respiratory difficulties are common. Occasionally, the animal vomits, but this is not an invariable characteristic of tympany. If vomiting occurs, the distressed animal may inhale its vomitus. Radiographs usually reveal gas-filled loops of intestine.

Treatment for bloating consists of administering a gas-lysing agent such as simethicone. There are several non-prescription liquid products which contain this agent and each can easily be delivered via a stomach tube. As mentioned previously, Mylicon (Stuart) has proven particularly efficacious. Where deemed appropriate, a modest dose of neostigmine or physostigmine can be injected to gently stimulate intestinal motility and thus, help relieve the retention of gas. Encouraging the affected animal to exercise may aid in the passage of excessive intestinal gas. The owner or keeper must be informed of the causes of this condition and should be encouraged to alter the diet so that bloating does not recur.

ARTICULAR AND PERIARTICULAR PSEUDOGOUT

A rare disorder of calcium metabolism mimics the clinical manifestations of articular and periarticular gout. This disorder is due to an abnormal deposition of calcium hydroxyapatite crystals within and around joint tissues. It was first reported in a two-year-old female red-eared slider turtle, *Trachemys scripta elegans* (Frye and Dutra, 1976). Although rare, this condition is included in this manual because it is diet-induced and is unlikely to occur spontaneously in reptiles feeding upon natural diets in the wild. This turtle had been fed a diet consisting almost exclusively of cooked shrimp.

All four legs were involved, and numerous internal lesions were revealed by whole-body radiography. The postmortem examination of the turtle revealed multifocal accumulations of cream-colored gritty material which surrounded the joint capsules. The visceral organs were unremarkable; specifically, the sac surrounding the heart was free of extraneous material. Jackson and Cooper (1981) reported a case of pseudogout in a *Uromastix* lizard in which multiple accumulations of hydroxyapatite were found around the joints of the pelvic and brachial girdles.

TOXIC PLANT POISONING

There are many species of mildly irritating to profoundly toxic wild and cultivated plants that are occasionally ingested by captive reptiles, particularly terrestrial tortoises. Table 7 is a partial list of the more common species that have been implicated in plant intoxications.

Apple and pear seeds and the inner pit seeds of apricot, peach, plum and nectarine, while toxic, are rarely a cause for intoxication in reptiles. Apple and pear seeds are passed in the feces undigested; the other seeds are well protected by hard pits. Both contain prussic (hydrocyanic) acid and if the seeds are sufficiently crushed at the time of ingestion they can induce fatal intoxication.

Occasionally, rhubarb and other oxalate-containing plants or antifreeze solutions containing ethylene glycol are ingested; unfortunately, the diagnosis is not made until calcium oxalate crystals are found in tissue specimens obtained postmortem.

When diagnosed antemortem, the treatment for intoxication is usally symptomatic and is directed initially toward combatting the clinical effects of the intoxicant and removing as

Table 7
Toxic Plants

Plant name	Toxic Portion(s)
Acokanthera	Flowers and fruit
Aconite (monk's hood)	Roots, flowers, leaves and seeds
Amaryllis	Bulb, stem, flower parts
Amsinckia (tarweed)	Foliage and seeds
Anemone	Leaves, flowers
Apple (seeds only)	Seeds (only if crushed)
Apricot (seeds only)	Inner seeds
Autumn crocus	Bulbs
Avocado	Foliage
Azalea	Foliage, flowers
Baneberry	Foliage, fruits
Beach pea	Foliage, peas, and pods
Betal nut palm	All parts
Belladonna	Berries and others parts
Bittersweet	Berries
Bird of paradise	Foliage, flowers, seed pods
Black locust	Bark, sprouts, and foliage
Bleeding heart	Foliage, flowers, and roots
Bloodroot	All parts
Bluebonnet	Foliage and flowers
Bottlebrush	Flower parts
Boxwood	Foliage and twigs
Buckeye horse chestnut	Sprouts and nuts
Buttercup	All parts
Caladium	All parts
Calla lily	All parts
Cardinal flower	All parts
Carolina jessamine	Foliage, flowers, and sap
Casava	Roots
Castor bean	Uncooked beans
Chalice or trumpet vine	All parts
Cherry (inner seeds only)	Inner pit seeds
Cherry laurel	Foliage and flowers
China berry tree	Berries
Christmas berry	Berries
Christmas cactus (*Euphorbia*)	Entire plant
Christmas rose	Foliage and flowers
Columbine	Foliage, flowers, seeds
Common privet	Foliage and berries

Continued

Table 7 (Continued)

Plant name	Toxic Portion(s)
Coral plant	All parts
Crocus	Bulbs
Croton	Foliage, shoots
Cyclamen	Foliage, stems, and flowers
Daffodil	Bulbs, foliage, flowers, and pods
Daphne	Berries
Death camus	All parts are toxic; esp. roots
Deadly nightshade	Foliage, unripe fruit, sprouts
Delphinium	Bulbs, foliage, flowers and seeds
Destroying angel (death cap)	All parts of the mushroom
Dogwood	Fruit mildly toxic
Dumb cane (*Dieffenbachia*)	Foliage
Eggplant	Foliage only
Elderberry	Leaves, bark, and shoots
Elephant ear (taro)	Foliage
English ivy	Esp. berries
Euphorbia (spurge(s))	Foliage, flowers, latex-like sap
False hellebore	All parts
Fiddle neck (*Senecio*)	All parts
Fly agaric (amanita, deathcap)	All parts (cap and stem)
Four o'clock	All parts
Foxglove	Foliage and flowers
Gelsemium	All parts
Golden chain	Seeds and pods
Hemlock roots (water & poison)	All parts
Henbane	All parts
Holly, English and American	Foliage and berries
Horse chestnut	All parts
Horsetail reed (*Equisetum*)	All parts
Hyacinth	Bulbs, foliage and flowers
Hydrangea	All parts
Impatiens (touch-me-not)	All parts
Iris (flags)	Bulbs and roots, foliage & flowers
Ivy (all forms)	foliage and fruit
Jack-in-the-pulpit	Roots are mildly toxic
Jasmine	Foliage and flowers, esp. nectar
Jasmine, star	Foliage, flowers
Jatropha	Seeds and oily sap
Jerusalem cherry	Foliage and fruits
Jessamine	Berries

Continued

76 A Practical Guide for Feeding Captive Reptiles

Table 7 (Continued)

Plant name	Toxic Portion(s)
Jimson weed (thorn apple)	Foliage, flowers and pods
Johnson grass, wilted	All parts
Lambkill (sheep laurel)	Foliage
Lantana camara	Foliage, flowers, and esp. berries
Larkspur	Entire young plant; seeds & pods
Laurel	All parts are toxic
Lily of the valley	Foliage and flowers
Lobelia	All parts
Locoweed	All parts
Locust(s)	All parts
Lupine	Esp. seeds and pods, foliage
Machineel	All parts
Marijuana	All parts
May apple	Fruit
Mescal	All parts may be toxic
Milk weed	Foliage
Mistletoe	Foliage and berries
Moccasin flower	Foliage and flowers
Monkshood	Entire plant, including roots
Moonseed	Berries
Morning glory	Foliage, flowers, and seeds
Mountain laurel	Young leaves and shoots
Mushrooms (some wild forms)	Entire cap and stem
Narcissus	Bulb, flowers
Natal cherry	Berries, foliage
Nectarine (inner seed only)	Only inner pit seeds
Nicotine, tree, bush, flowering	Foliage and flowers
Nightshades	All parts, esp. unripe fruits
Oak trees	Leaves and acorns
Oleander	Foliage, stems, and flowers
Peach (inner seed only)	Inner pit seeds
Pear seeds	Seeds (only if crushed)
Pennyroyal	Foliage and flowers
Peony	Foliage and flowers
Periwinkle	All parts
Philodendrons, some species	All parts
Pinks	All parts
Plum seeds	Inner seeds; foliage can be toxic
Poinsettia	Foliage, flowers and latex sap
Poison hemlock	Foliage and seeds

Continued

Table 7 (Continued)

Plant name	Toxic Portion(s)
Poison ivy	Foliage and fruit
Poison oak	Foliage and fruit
Poison sumac	Foliage and fruit
Pokewood or pokeberry	Roots, fruit
Poppy (except California)	All parts
Potato	Raw foliage and sprouts ("eyes")
Privet	Berries
Redwood	Resinoids leached when wood is wet
Rhubarb	Uncooked foliage and stems
Rhododendron	Foliage and flowers
Rosary pea	Foliage, flowers and peapods
Rosemary	Foliage in some species
Russian thistle	Foliage and flowering parts
Sage	Foliage in some species
Salmonberry	Foliage and fruit
Scarlet pimpernel	Foliage, flowers and fruit
Scotch broom	Seeds
Senecio ("fiddle neck")	All parts
Skunk cabbage	Roots
Snapdragon	Foliage and flowers
Spanish bayonet	Foliage and flowers
Squirrel corn	Foliage, flowering parts, and seeds
Sudan grass, wilted	All parts
Star of Bethlehem	Foliage and flowering parts
Sundew	Foliage
Sweetpea	Stems
Tansy	Foliage and flowers
Taro (elephant ears)	Foliage
Tarweed	Foliage and seeds
Tiger lily	Foliage, flowers, and seed pods
Toad flax	Foliage
Tomato plant	Foliage and vines
Toyon berry	Berries
Tree of heaven	Foliage and flowering parts
Trillium	Foliage
Trumpet vine	All parts
Tulip	Bulb, foliage, and flowering parts
Venus flytrap	Foliage and funnel flowering parts
Verbena	Foliage and flowers

Continued

Table 7 (Continued)

Plant name	Toxic Portion(s)
Vetch (several forms)	Seeds and pods
Virginia creeper	Foliage and seed pods
Water hemlock	Roots and foliage
Wild parsnip	Underground roots and foliage
Wisteria	Foliage, seeds, and pods
Yellow star thistle	Foliage and flowering parts
Yew	Foliage

Modified from a list compiled by the *International Turtle and Tortoise Journal* May–June, 1969, and a compilation by the San Diego Turtle and Tortoise Society, published in the *Tortuga Gazette* January, 1982.

much of the material as possible to prevent further absorption from the alimentary system. To accomplish this efficiently, warm water gastric lavage (stomach washes) and enemas may be administered. Because of the common cloaca into which the gastrointestinal and genitourinary tracts empty, great care must be taken when introducing enema solutions into the cloacal vent. This treatment should be reserved for those clinical situations when alternative means of relieving colonic impaction or intoxication are deemed inappropriate. Fluid must never be forced into the cloaca under pressure because it may transmit disease causing organisms from the cloacal vault into the terminal genitourinary tracts. Supportive medical management with injections of calcium gluconate, atropine sulfate, diuretics, and fluid therapy is usually indicated.

Finally, a rather common yet very mild form of intoxication is observed in tortoises given access to rotting, overripe, and fermenting fruit that has fallen from trees within the confines of the reptiles' enclosure or yard. The author's luxuriantly fecund apple, peach, and nectarine trees have supplied the substrate for many a heady, drunken binge by our tortoises. Aside from staggering around and bumping into each other (and stationary objects) for a few hours, the animals are none the worse for their lack of sobriety.

SPECIAL NUTRITION-RELATED BEHAVIORS

Introduction

Some reptile eating habits, while alarming to their owners, may not be abnormal. For instance, some reptiles will attack and eat their own kind. Others will actively seek out and ingest rocks, pebbles, and sand from the confines of their captive environment. In some of these animals, this behavior is entirely normal because these stones serve to help digest the rough food upon which these animals subsist. In others, the stones serve as ballast and help maintain a neutral or negative buoyancy so that the animal does not have to expend much energy in staying on the bottom of a body of water.

Other behaviors, such as the eating of one's own eggs, feces, or even glassy silica-rich sponges have been reported in both wild and captive reptiles and often can be controlled by proper management.

Cannibalism

Some degree of cannibalism may be observed in many snakes, lizards, crocodilians and even some turtles. In many cases, this behavior is associated and even triggered by a "feeding frenzy" as the animals in a tank or cage snap at food items that are offered. In other instances, overt attacking and eating of one's own species occurs. This behavior is common in the garter snakes, *Thamnophis* sp. and has been reported in the Great Basin gopher snake, *Pituophis melanoleucus deserticola* by Zaworski (1990). The latter occurred under unusual circumstances and may not be typical of the species. The provision of suitable hiding boxes, hollow logs, or other refuges in the cages in which several snakes or lizards are being housed will greatly reduce the incidence of cannibalism.

Dermatophagy

It is not unusual for many lizard species to ingest part of their entire molted epidermis (Bustard & Maderson, 1965). Dermatophagy has also been reported in at least three species of snakes (Bustard & Maderson, 1965; Groves, 1964; Groves and Altimari, 1977; Groves & Groves, 1972; Keown, 1973). This behavior is entirely normal in several species of geckos and should not be a cause for concern. The molted old epidermis represents a suitable source of sulfur-containing amino acid rich protein which is, therefore, recycled.

Autoovophagy

Occasionally a reptile kept in captivity will consume its own eggs. This abnormal behavior has been witnessed in the emerald tree boa, *Corallus enhydris enhydris*, (Miller, T.J., 1983), the ball python, *Python regius*, and several species of geckos. One means for lessening this activity and resultant loss of eggs is to provide a sufficiently large cage and adequate hiding places and/or litter in which a female reptile may deposit her eggs.

Coprophagy

The consumption of stools is unusual in most reptiles. When it does occur, the feces eaten are usually not those of the animal that is ingesting them. The author has witnessed avid consumption of dog and tortoise stools by terrestrial chelonians, particularly the California desert tortoise, *Xerobates (Gopherus) agassizi* and the Texas tortoise, *X. (Gopherus) berlandieri*. Banks (1984) reported coprophagy in the cobra, *Naja melanoleuca*, and Peterson (1980) described this behavior in a garter snake.

Most, but not all, intestinal roundworms, tapeworms and flukes are rather host-specific. However, there is an obvious opportunity for the horizontal transmission of disease causing bacteria, fungi, viruses, protozoa, and metazoa from an infected or infested animal to cage mates, particularly if they are of the same or closely related species. Moreover, many parasitic metazoa are characterized by their direct or indirect multihost life cycles and may be furnished with the required host when an animal consumes the stools of another. This behavior should be discouraged by strict attention to hygiene and avoidance of overcrowded conditions.

Gastrointestinal Silicosis in Marine Turtles

Sea turtles, especially the hawksbill, *Eretmochelys imbricata*, are known for their habit of eating coral reef sponges; in one study, sponges comprised 95.3% of the total dry mass of all food items in the digestive tract of 61 hawksbill turtles from seven Caribbean countries (Meylan, 1988). These invertebrates contain substantial amounts of silica and are known to be toxic to most other vertebrates. In her paper, Meylan discussed the effect of the hawksbill turtles' sponge eating upon existing coral reef communities and pointed out the apparent ability of the hawksbill, common green, *Chelonia mydas mydas*, and loggerhead, *Carretta carretta* turtles to not only survive, but also often specialize on such an otherwise unpalatable silica-containing diet. Some of these sea turtles also eat calcareous corals. At Steinhart Aquarium in San Francisco, we experienced substantial loss to the artificial coral reef displays that were part of the sea turtle enclosures. These displays were constructed from sculptured concrete and epoxy-impregnated fiberglass. Hawksbill turtles chronically ingested these materials but did not appear to suffer ill effects from this depredation upon their captive environment.

Lithophagy and Geophagy

It is a common finding to observe the presence of pebbles and even rather large stones in the gastrointestinal tracts of crocodilians and terrestrial, aquatic, or semiaquatic turtles. Often sand or small pieces of rock are found in the intestines of terrestrial lizards. Several survey papers have been written which deal with the apparent voluntary ingestion of earth and stones by a variety of reptiles (Kramer, 1973; Peaker 1969; Rhodin 1974; Skorepa 1966; and Sokol 1971) Some authorities have suggested that this behavior is related to buoyancy control in aquatic or semiaquatic species and also to maceration and/or early processing of scabrous plant material or bony prey. Both requirements appear to be met by lithophagy where particulate size of the ingested stones or pebbles would tend to keep these foreign bodies *in situ* within the stomach cavity.

However, when smaller particles of earth, sand or gravel are ingested as a result of being accidentally carried as contaminants of moist or sticky food items, the result is often variable blockage of the intestinal tract. These blockages may lead to gross obstruction and dysfunction. Some of these loam, sand, gravel and/or rock obstructions can be relieved by conservative measures employing laxatives, enemas and/or warm water soaks. Some of the larger rocks can be grasped with an appropriate long-jawed forceps and removed via the esophagus. Sometimes intestinal surgery is necessary. As is the case with many other husbandry-related problems, prevention has much merit.

Sand or gravel litter should be avoided whenever possible, and food items should not be fed in such a manner that they are likely to attract sand or gravel onto their surfaces. One material that provides both a safe and highly absorbant litter—and at the same time is highly nutritious when ingested by herbivorous reptiles—is pelleted alfalfa rabbit chow. It is inexpensive and readily available in pet and feed stores. It must be changed whenever it becomes soiled or moist.

3

Sources of Reptile Food

INTRODUCTION

While it may, at times, appear to be an almost unsurmountable task to obtain a varied and nutritional diet for many reptilian species, the truth is that many chelonians and lizards may be fed well balanced diets with herbaceous materials obtained from the left overs of the nearest green grocer, supermarket, and/or commercial florist shop. As soon as fruits and vegetables develop blemishes or become slightly wilted, they are consigned to the trash receptacle. Because of the constraints imposed upon farmers by the Department of Agriculture, long-acting toxic pesticides are not used on food plants destined for human or domestic animal consumption. Commercial flowers are most often grown in greenhouses where pesticides are unnecessary. Another reason for utilizing slightly "distressed" commercial produce is that it simply makes good sense to recycle this otherwise culled refuse or "garbage" and put it to a far more valuable use as animal food. Rare indeed is the produce manager who, once he knew to whom it was going to be fed, has refused a request for such material. Similarly, florists have no use for slightly wilted rose petals and are universally delighted to have an outlet for them, particularly if they know that those petals will be fed to animals.

Excellent green fodder may be grown specifically for captive reptiles in simple flower pots and/or window planter boxes. To be sure, one's neighbors may perceive a window box displaying a luxurient crop of dandelions as a bit tacky, but such mini-farms furnish an outstanding source of fresh greens and blossoms. Hydroponic systems can produce abundant young, tender grasses and alfalfa, as necessary. Bean, seed, and pulse sprouts are very easily grown in any volume dictated by the size of a particular animal collection; they also make good human food! A brief discussion of each of these techniques follows this section.

On the following two pages numerous plant materials are listed that are easily obtainable. Each of the items listed is nutritious and palatable. For best results, several items from this list should be combined to formulate a vegetarian diet that will be readily acceptable and will furnish the macro- and micro-nutrient requirements for herbivorous terrestrial reptiles. Some of these items can be grown easily by the hobbyist; others can be obtained either from an animal feed store or human grocery. With some ingenuity, even the most exotic reptile can be fed a ration that not only meets its dietary preferences, but one that is nutritionally and environmentally sound and, in some cases, can be shared with the human inhabitants of the household.

Examples of Plant Materials that Are Safe and Nutritious

Alfalfa: fresh, sun-cured hay, dried leaves, pellets, meal
Apple: fresh, with peel, sliced or grated (discard core and seeds)
Barley: freshly sprouted seeds, freshly grown leaves, sun-cured hay
Beans (several edible varieties): fresh leaves and stems, fruit
Bean sprouts (azuki, black-eyed, garbanzo, lentil, mung, pea, etc.): fresh leaves, stems, blossoms, fruit
Beet: tops, stems, flowers, grated roots
Berseem (Egyptian clover): leaves, sun-cured hay
Buffalo grass (*Bulbilis dactyloides*): hay
Cabbage family (kale, napa, broccoli, Brussels sprouts): do not feed to excess.
Cactus: flowers, prickly pears, tender young cactus pads
Carrot: leaves, grated root
Clover (Ladino, Alsike, etc.): fresh, sun-cured hay

Collards: fresh green leaves, flowers
Cotton: leaves, dried or fresh
Cowpea: sun-cured hay, leaves
Crucifers: bok choy, etc.
Dandelion: leaves and stems, flowers, fresh or dried
Dicondra: fresh or sun-cured hay
Eugenia: fresh leaves, fruits
Figs: fresh
Grass clippings: freshly mowed or sun-cured
Hibiscus: leaves, flowers, fresh pods
Kudzu: sun-cured hay
Lespedeza: sun-cured hay, leaves
Millet: leaves, sun-cured hay
Mint: sun-cured hay
Mixed vegetables: frozen, thawed
Mulberry: freshly picked tender leaves, fruit
Mustard: fresh green leaves, flowers
Nasturtium: leaves, stems, flowers
Okra: fresh, chopped, tender leaves, flowers
Pea: fresh pods, sun-cured hay
Pear: fresh, cut or grated (discard core and seeds)
Peavine: sun-cured hay
Peanut: sun-cured hay with or without nuts
Pelleted commercial chows (Purina, Wayne, etc.) for guinea pigs and rabbits can be fed *ad lib*; those formulated for horses, goats, dogs, cats or monkeys, etc. *Should not be fed in excess.*
Rape: fresh leaves, sun-cured hay
Rutabaga: freshly grated root
Saltbush (winter range): sun-cured hay
Soybean: fresh leaves or sun-cured hay
Squash: freshly grated flesh, blossoms, tender leaves
Stone fruits: peach, nectarine, apricot, plum, etc.
Sunflower: seeds (unsalted)
Timothy: sun-cured hay
Tofu soybean cake
Triticale: freshly sprouted seeds, sun-cured hay
Turnip: fresh leaves, grated root
Vetch: sun-cured hay
Wheat (soft wheat berries): freshly sprouted, hydroponically grown

Table 9
Food Values

Food	Measure	Vitamins A units	B-1 (mg)	B-2 (mg)	C (mg)	Minerals Calc. (mg)	Phos. (mg)	Iron (mg)	Other Prot. (gm)
Apple	1 small	90	.360	.050	6	7	12	0.3	0
Apricot #	3 med.	7,500	.033	.100	4	13	24	0.6	1
Asparagus	8 stks	1,100	.360	.065	20	21	40	1.0	2
Avocado	½ med	500	.120	.137	9	44	42	6.3	2
Banana	1 med.	300	.045	.087	10	8	28	0.6	1
Beans, green*	¾ cup	950	.060	.100	8	55	50	1.1	2
Beet greens*	½ cup	22,000	.100	.500	50	94	40	3.2	2
Beets	½ cup	50	.041	.037	8	28	42	2.8	2
Blackberries	¾ cup	300	.025	.030	3	32	32	0.9	0
Blueberries	¾ cup	35	.045	.031	11	25	20	0.9	0
Broccoli flr	¾ cup	6,500	.120	.350	65	64	105	1.3	2
Broccoli leaf	¾ cup	30,000	.120	.687	90	262	67	2.3	3
Broccoli stem	¾ cup	2,000	—	.187	—	83	35	1.1	2
Brussels spr	¾ cup	400	.180	.090	130	27	121	2.1	4
Cabbage (1)	1 cup	0	.780	.075	50	46	34	2.0	2
Cabbage (2)	1 cup	160	.090	.150	50	429	72	2.8	2
Cabbage (3)	1 cup	5,000	.036	.462	50	400	72	2.5	2
Canteloupe	½ small	900	.090	.100	50	32	30	0.5	1
Carrots (4)	½ cup	4,500	.070	.075	5	45	41	0.6	1
Cauliflower	¾ cup	10	.085	.090	75	122	60	0.9	2
Celery (5)	4 stlk	20	.030	.015	5	78	46	0.5	1
Celery, grn	4 stlk	640	.030	.045	7	98	46	0.8	1
Celery root	½ cup	—	—	—	2	47	71	0.8	3
Chard, lvs*	½ cup	15,000	.450	.165	37	150	50	3.1	2
Cherries #	12 lrg	259	.051	—	12	19	30	0.4	1
Collards*	½ cup	6,300	.130	—	70	207	75	3.4	3
Corn on cob	1 med.	860	.209	.055	8	8	103	0.4	3
Cucumber	1 med.	35	.060	.054	12	10	21	0.3	1
Dandelion grn*	½ cup	20,000	.190	.270	100	84	35	0.6	3
Eggplant	½ cup	70	.042	.036	10	11	31	0.5	1
Endive	10 stks	15,000	.058	.072	20	104	39	1.2	1
Grapefruit	½ med.	20	.070	.060	45	21	20	0.2	0
Grapes	1 sm bnch	25	.030	.024	3	19	35	0.7	1
Guavas	1	200	.156	.105	125	15	16	3.0	1
Honeydew mel	¼ med.	100	—	—	90	—	—	—	0
Huckleberry	½ cup	100	.045	.021	8	25	20	0.2	1
Kale*	½ cup	20,000	.189	.570	96	195	67	2.5	4
Kohlrabi	½ cup	—	.030	.120	50	195	60	0.7	2

Table 9 (Continued)

Food	Measure	Vitamins				Minerals			Other
		A units	B-1 (mg)	B-2 (mg)	C (mg)	Calc. (mg)	Phos. (mg)	Iron (mg)	Prot. (gm)
Leeks	½ cup	20	.150	—	24	58	56	0.6	2
Lettuce, grn	10 lvs	2,000	.075	.150	7	49	28	1.5	1
Lettuce, wht	¼ head	125	.051	.062	5	17	40	0.5	1
Mushrooms (6)	¾ cup	0	.160	.070	2	14	98	0.7	4
Mustard gr.	½ cup	11,000	.138	.450	126	291	84	9.1	2
Okra	½ cup	440	.126	—	17	72	62	2.1	2
Onions, frsh	4 med	60	.042	.125	7	41	47	0.4	1
Orange	1 med.	190	.090	.075	50	44	18	0.4	0
Parsley	½ cup	8,000	.057	—	70	23	15	9.6	20
Parsnips (6)	½ cup	100	.120	—	40	60	76	1.7	2
Peaches, wht	3 halves#	100	.025	.065	6	10	19	0.2	1
Peaches, yel	1 lrg#	1,000	.025	.065	9	10	19	0.3	1
Pear	1 med.	17	.030	.060	4	15	18	0.3	0
Peas, fresh*	½ cup	1,500	.390	.250	20	28	127	0.2	7
Persimmon (7)	1 lrg	1,600	—	—	40	22	21	0.2	2
Pineapple†	⅔	30	.100	.025	38	8	26	0.2	0
Plums	3 med.	130	.120	.056	5	20	27	0.5	1
Potato, swt	1 med.	3,600	.155	.150	25	19	45	0.9	3
Potato, wht	1 med.	0	.220	.075	33	13	53	1.5	3
Potato, yam	1 med.	5,000	.180	.360	6	44	50	1.1	2
Pumpkin	½ cup	2,500	.056	.057	8	23	50	0.9	1
Radishes	15 lg	0	.030	.054	25	21	29	0.9	1
Raspberries	½ cup	260	.021	—	30	41	38	0.8	1
Rutabaga	¾ cup	25	.075	.120	26	74	56	0.7	1
Spinach*	½ cup	11,000	.090	.312	30	78	46	2.5	2
Squash, hubd	½ cup	4,000	.050	.075	3	19	15	0.5	1
Squash, sum.	½ cup	1,000	.040	.050	3	18	15	0.3	1
Strawberries	½ cup	100	.025	—	50	34	28	0.6	1
Tangerine	2 med.	300	.120	.054	48	42	17	0.2	1
Tomatoes	1 med.	1,500	.100	.050	25	11	29	0.4	1
Turnips*	½ cup	0	.062	.062	22	56	47	0.5	1
Turnip grns	½ cup	11,000	.060	.045	130	347	49	3.4	2
Watercress	¾ cup	1,250	.030	.090	15	40	11	0.8	0
Watermelon	1 med sl	450	.180	.084	22	33	9	0.6	0

(1) inside white leaves; (2) outside green leaves; (3) Chinese; (4) diced; (5) bleached; (6) domestic; (7) Japanese; #, pitted; *, cooked; †, fresh.
Modified from *International Turtle and Tortoise Society Journal*, August/September/October, 1970.

HYDROPONIC CULTIVATION OF GRASSES

With currently available hydroponic technology, high quality grasses can be readily produced in a limited space, without regard for climatic conditions and without the need for soil.

Several zoological gardens feed diets made from one or more hydroponically cultivated grasses (usually barley, wheat, oats and/or triticale) to their herbivorous animals. This technique has several distinct advantages over obtaining fresh green fodders from outside vendors: the initial capital expenditure is usually within the budgetary limits of most professional collections; the labor expenses required to seed, care for, and harvest each crop are modest; the quality and quantity of the grasses can be easily monitored and controlled; the quantity can be increased or decreased with little effort; the time required from seeding to harvest is quite brief, measuring only about one week; and the produce is pesticide- and parasite-free. The compact hydroponic growing units offer an opportunity to assure an economical and consistent supply of nutritious, highly palatable green feed.

Production models of these self-contained growing environments are, apparently, no longer manufactured, but it is possible to design and construct suitable units from off-the-shelf materials available at most building construction suppliers. One needs to provide a relatively closed environment in which the humidity and temperature can be controlled; an automatic illumination system employing fluorescent lighting of an appropriate wave length (examples are Sylvania Gro-Lux F20T-12/GRO, 20 watt, 61 cm; Sylvania Gro-Lux F40/GRO, 40 watt, 122 cm; Westinghouse Agrolite F20T12/AGRO, 20 watt, 61 cm; and Westinghouse Econ-O-Watt F40CW/RS/EW-II; 34 watt); growing trays in which seeds can be sown; and an automatic time-controlled submersible pump to distribute the growing medium. A rack arrangement is used to suspend the growing trays. Two or three times a day, the pump distributes a balanced growing solution to the growing trays; these trays are self-emptying through overflow standpipes so that they cannot overflow. After the pumps are turned off, any surplus growth medium is returned to a sump or other suitable container in which the pump is submersed. The photoperiod is controlled by a time clock which activates the light sources several times a day.

Smaller earthless growing units are available from some nursery dealers, or they can be made at home. One can construct several experimental models using a large diameter polyvinylchloride irrigation pipe split lengthwise to hold pea gravel sized tumbled volcanic pumice, and small submersible pumps to circulate the growth medium. These experimental units have some advantages over some of the commercially produced models and are substantially less expensive.

The amateur herpetologist or professional who has a relatively small collection may wish to build a small pilot model in which a more modest amount of fresh food can be grown. An excellent source of information, plans, and growing supplies is Hydro-Fresh Farm, P.O. Box 511, San Martin, CA 95046. The balanced growth medium is available from this firm. In addition, local, state, and federal agricultural extension offices should be consulted for details of the various processes and/or plans. Small quantities of green feed can be produced in jars. (See next section.)

BEAN, PEA, LENTIL, AND SEED SPROUT CULTURE

Home-sprouted seeds and pulses (peas, beans, lentils, etc.) are excellent sources for macro and micronutrients in diets for herbivorous reptiles. The ease of producing these sprouts in varying quantities and the low cost of equipment and seed stock make the culture of these nutritious and desirable items attractive for the reptile collections of large institutions as well as more modest private amateur hobbyist collections. Furthermore, these sprouts are appropriate for human consumption.

The following pulses are particularly appropriate for sprouting: small mung beans, azuki beans, garbanzo beans (chickpeas), whole green or yellow peas, and lentils. The following seeds are well suited for sprouting: soft wheat berries, triticale, maize, whole unsalted sunflower seeds, millet, rapeseed, alfalfa, and barley. Small quantities of radish seeds may be added to mixtures to enhance the excellent flavor of the resulting young sprouts. Radish sprouts impart a tangy flavor and are readily accepted by reptiles. However, too many make the entire sprout crop too spicy for most animals (and humans).

All beans, peas, lentils, and seeds must be obtained from sources intended for human consumption, such as health food stores. Sprouted seeds and pulses sold for garden or farm use must not be fed to animals or consumed by humans because they are often treated with fungicides and insecticides that are highly toxic.

The only equipment required to produce small quantitites of fresh sprouts are 1 quart or larger glass canning jars fitted with outer rings holding disks of fiberglass or stainless steel screening mesh. Larger volumes of sprouts require correspondingly larger containers.

To induce sprouting, 1½ to 2 tablespoons of dry mixed seeds and pulses are placed into the container, and slightly tepid tap water is added to cover the seeds to a depth of approximately 10 cm. The seeds are allowed to soak overnight. As they soak, they absorb water and often swell to twice their original size. After approximately 12 hours of soaking, the water is poured off through the screened lid and the now softened mixture of seeds is allowed to germinate in a dark area. After approximately 36 hours, the first tiny sprouts appear. Each day, rinse the jars of sprouting seeds and pulses in cold water and thoroughly drain them through the screened lid. After 3 to 5 days, the sprouts are ready to harvest. At that time, rinse the sprouts in cold water, remove them from the jars, and use them immediately or store them in a refrigerator. Culturing the sprouts in total darkness is particularly important if some of them are to be consumed by humans because, as the sprouts mature, they develop chlorophyll which may impart a bitter flavor to the now green sprouts. Although this flavor may be objectionable to humans, it does not appear to affect the palatability to herbivorous reptiles and, in fact, actually improves the nutritional value of the sprouts.

Depending on the varieties and quantities grown, a few cents' worth of dry seeds and pulses provide over a liter of fresh sprouts in as few as four days, and they may be produced every month of the year. Pulses and alfalfa are excellent sources of high quality vegetable protein, vitamins, minerals (including calcium), and useful energy in the form of cellulose and other complex carbohydrates.

Fresh sprouts are especially valuable for treating tortoises and iguanas that have become accustomed to eating only iceberg lettuce, cabbage, or other nutritionally deficient vegetable items. The sprouts should be sprinkled as a top dressing over the animal's preferred leafy items so that the animal must eat its way through the sprouts in order to reach the lettuce or cabbage, thus ingesting both its preferred items and those more nutritionally sound items placed in its gustatorial pathway. At each daily feeding, the proportion of sprouts is increased in relation to that of the lettuce or cabbage. Dandelions, nasturtiums, grasses, rose petals, etc., can also be placed in the reptile's field of vision. Usually within two weeks or less, the animal may be sufficiently retrained to accept a far more nutritional and varied vegetable diet. Sprouts also are a superb means of increasing the caloric and other nutritional factors of a ration prior to placing an herbivorous animal into hibernation.

HIGH-FIBER DIET FOR HERBIVOROUS REPTILES

Dr. J. A. Sweet, Director of The Last Resort organization, working with threatened iguanid species, has developed a captive diet which has proven to be nutritionally sound and is well accepted by the animals to which it is fed. It is reproduced here with Dr. Sweet's permission:

- natural wheat bran cereal
- alfalfa pellets
- 9-grain bread
- alfalfa sprouts
- clover sprouts
- natural honey
- Superpreen®—Blair's Products, RHB Laboratories, Inc., 1640 East Edinger Avenue, Santa Ana, CA 92705
- vitamin B complex with folic acid

purified water
calcium carbonate powder

Equal portions of bran and alfalfa pellets are mixed with 0.5 kg of 9-grain bread and 0.25 kg each of alfalfa and clover sprouts. This mixture is moistened with 120 ml of honey and 720 ml of water. The dosage of Superpreen is 0.063 tsp (approximately 12 mg) per kg of body weight. Vitamin B complex supplement is added to the mixture at a rate that will yield a dosage of 50 mg/kg twice weekly. Calcium carbonate is administered at a rate of 200 mg/kg/week.

Dr. Sweet has raised several generations of *Iguana iguana* and *Cyclura* sp. on a diet of this formula in addition to mixed fruits and melons, mixed raw green and yellow/orange vegetables (yams, sweet potatoes, squash), grasses, legumes, flowers and buds (roses, dandelions, etc.), cactus pads, sprouted seeds, infant cereals, brown rice, cereal breads, and assorted cooked meats and eggs. She recommends that animal protein sources be fed only twice monthly. Immature lizards are fed *ad lib*, adults are fed three times weekly.

As mentioned earlier in this chapter, the gastrointestinal tract of most herbivorous reptiles is populated by a variety of symbiotic microorganisms. A naive herbivore should be fed gradually increasing amounts of concentrated diets for several weeks so that its intestinal microflora can become accustomed to a diet containing such complex carbohydrates as honey, cereal grains, and bread. Such an acclimatization period may require several weeks after a diet of mostly vegetables and fruit.

MEAT DIETS:
NOTES ON THE CULTURE OF PREY SPECIES

Introduction

Although crickets, mealworms, mulberry silk moth larvae ("silkworms") and earthworms may be purchased from pet, aquarium, biological suppliers, and fish bait stores, respectively, often it is more convenient and economical to raise these invertebrates at home or within the confines of a professional collection or colony. They are not difficult to raise and any surplus often may be traded to pet or aquarium stores for other items of interest.

These invertebrates are all easily cultured under controlled conditions (Frye, 1987, 1991). Growing such invertebrates has several advantages: the supply and quality can be controlled, and the likelihood of parasitic transmission via prey food items can be greatly reduced. Culturing some of these invertebrates does demand responsibility to see that the insects are not allowed to escape. For example, wax worms, if allowed to complete their metamorphosis, can prey upon the beeswax combs of domestic beehives. Similarly, crickets, mealworm beetles, and even genetically vestigial winged or wingless fruit flies can become a nuisance if they establish themselves outside the confines of their assigned culture containers.

Arthropods, particularly crickets and silkmoth larvae, can be dusted with ground calcium carbonate or Pet Cal (Beecham-Massengill) tablets to improve their calcium content just prior to feeding them to insectivorous reptiles. Similarly, mealworms may be calcium enriched by adding ground alfalfa pellets, Pet Cal tablets, dry dog kibble, monkey chow, or poultry mash to their growing medium.

Mulberry Silk Moth Larvae

Long cultured in the Orient for their valuable silk cocoons, the mulberry silk moth, *Bombyx mori*, can be an extremely useful invertebrate for feeding a wide variety of large mygalomorph (orthognath) spiders, scorpions, solifugids, amphibians and many carnivorous lizards and bird species. Captive insectivorus bats, moles, and shrews also find these plump grub-like larvae to their liking.

Culture

The silkworm larvae of the domestic mulberry silk moth should be kept in screened wooden or plastic cages whose volume depends upon the number of larvae to be maintained at any one time. For a population of four dozen larvae, a one to one-and-a-half cubic foot enclosure is sufficient. At first, this may appear to be a rather commodious arrangement, but because these larvae grow so rapidly, they will soon begin to crowd each other. When provided with an adequate supply of freshly picked or properly frozen mulberry leaves, these soft-bodied larvae grow rapidly and can be harvested as required for food prey. Clean paper toweling or dried mulberry leaves serve well as cage litter and should be changed when they become soiled. Routine hygiene will help prevent an accumulation of feces and uneaten plant material. Actively feeding larvae usually do not seek escape from their culture containers as long as fresh mulberry leaves are provided, but as soon as they are mature enough to spin their silken cocoons, they seek vertical surfaces and will escape if not limited in their travels by a screened enclosure.

Freshly deposited or properly stored fertile eggs require 7–10 days to hatch. The fertilized eggs require a period of refrigeration to interrupt the resting phase, or diapause, that under natural conditions prevents the embryonic development of the larvae during winter when a supply of mulberry leaves is not available. Under artificial culture, the eggs are placed in a refrigerator for at least 2 weeks to accelerate the diapause. Under natural conditions, this diapause lasts about 5–6 months. Fertile silk moth eggs can be stored for about a year as long as they are not permitted to warm to room temperature. They must not be frozen because this kills the embryonic larvae. The eggs are removed from the refrigerator and are placed onto freshly picked or growing mulberry leaves still attached to a tree. The eggs hatch in about 3–5 days and the tiny larvae begin to feed immediately.

The silk moth larvae feed actively for 25–30 days during which time they multiply their mass by several orders of magnitude. The fecal pellets of these larvae are rich in plant nutrients and make excellent fertilizer for house plants. Then the larvae cease eating and begin to spin their cocoons, which usually are completed in 1–2 days. Once the larvae are encased in their cocoons, metamorphosis begins and is usually completed in 10–18 days. The adult silk moths emerge from their cocoons, mate, deposit their several hundred eggs, and soon die. This last stage of their brief life cycle lasts from 3–5 days in females and about 1–2 days in males.

Nutrition

Silk moths are highly adapted to a restricted diet of mulberry leaves and those of closely related trees. Whether these leaves are harvested from fruited or fruitless mulberry trees does not appear to make any significant difference in acceptance by the larvae as long as the leaves are free from pesticide residues. Some of the heavier twigs or branches of the mulberry tree should be included so that the larvae have a place to distribute themselves, rest, and/or spin their silk-wound cocoons. Cooper (1961) suggests Osage orange leaves as an acceptable substitute for mulberry trees. Frye (1987) found that the season for culturing silk moth larvae can be extended beyond the normal growing season of mulberry trees by freezing and storing lightly blanched leaves. The individual leaves are blanched in boiling water for about 12–15 seconds, patted dry with either paper or clean cotton bath toweling, and then placed into Ziploc® freezer storage bags from which excess air has been expressed. An alternative to the use of closeable reusable bags is to vacuum pack the blanched leaves in food storage pouches and keep them from becoming freezer-burned and dessicated. It is important that the leaves be blanched prior to freezing to inactivate enzymes within the leaf tissue. This will help retain the freshness in the leaves. Once the leaves have been thawed for a few minutes, they should be patted dry and fed immediately. The larvae prefer freshly picked leaves, but they will consume previously frozen leaves if they have been properly prepared.

To furnish additional calcium, the larvae can be dusted with calcium carbonate or some similar calcium containing product. Alternatively, they can be injected with calcium gluconate or lactate immediately before feeding them to predators. The author found that silk moth larvae will imbibe calcium gluconate if it is sprayed onto fresh leaves. However, these larvae should be fed to predator reptiles within a few hours because the treated larvae begin to display distress within about 6–10 hours and die within 24 hours after they have swallowed the concentrated calcium supplement; reptiles feeding upon even obviously dyspeptic larvae remain unaffected by their prey's distress.

Water

When fed fresh mulberry leaves, the silk moth larvae obtain required moisture from their leafy diet. When frozen leaves are fed, additional water can be provided by *lightly* misting the leaves once or twice daily, but the leaves must not be allowed to become soggy because the larvae will refuse to eat them and also because this will promote the growth of harmful fungi.

Reproduction

The mulberry silk moths mate soon after they emerge from their cocoons. It is for this reason that a few larvae should be allowed to complete their metamorphosis to ensure the next generation. It is important to select as breeders only those larvae that appear large and vigorous and to cull any moths with obvious physical defects. If a supply of mulberry leaves is available or if one can be assured of an unlimited supply of frozen leaves, a nearly year-round culture of larvae can be provided.

Soon after mating, the female moths deposit their eggs onto mulberry twigs and leaves. After mating the adult male moths die, but the females may survive for several days during which they mate repeatedly and deposit clutches numbering several hundred eggs. Fertile eggs soon darken; infertile eggs remain a creamy pale yellow and should be discarded.

Because of the high lipid content of their preovulatory ovaries, the nutritional value of the female moths is not as great after they have deposited their eggs. However, they still contain valuable nutrients.

Silk moth larvae have been analyzed for their chemical composition under two separate sets of circumstances: as plain and as calcium enriched individuals. Tables 10A and 10B list the chemical compositions of these two experimental populations of silk moth larvae. As can be seen, these soft grub-like

Table 10A
Chemical Composition of Silk Moth Larvae

Component	Silkworms[c]	Silkworms[d]	Mulberry Leaves
Moisture	76.3	69.9	63.1
Nitrogen	10.35	10.03	2.68
Protein[a]	64.7	62.7	16.8
Fat	20.83	14.15	3.49
Calcium	0.21	0.24	0.69
Phosphorus	0.54	0.57	0.17
Magnesium	0.24	0.26	0.70
Energy[b]	5.74	5.69	ND[e]

From Frye and Calvert, 1989a. Consitutents are expressed as % of dry matter.
[a]Nitrogen times 6.25.
[b]Heat of combustion.
[c]Analyzed with intestinal contents intact.
[d]Analyzed with intestinal contents removed.
[e]ND = not determined.

Table 10B
Chemical Composition of Calcium-Enriched Silk Moth Larvae

Component	Silkworms[c]	Silkworms[d]	Silkworms[e]	Controls
Moisture	65.13	65.04	61.49	62.07
Nitrogen	9.19	7.86	8.33	9.46
Protein[a]	57.40	49.10	52.10	59.10
Fat	12.89	11.35	9.02	10.54
Calcium	1.32	2.76	2.17	0.43
Phosphorus	0.69	1.12	0.78	0.91
Energy[b]	3.50	2.99	2.90	4.42

Fron Frye and Calvert, 1989b. Constituents are expressed as percentage of dry matter.
[a]Nitrogen times 6.25.
[b]Expressed on a dry matter basis.
[c]Larvae plus topical Cal Quick.
[d]Larvae plus topical Cal Quick plus Vionate Powder.
[e]Larvae fed 23% calcium gluconate immediately before analysis.

insect larvae possess a number of distinct advantages over meal beetle larvae and crickets. Also, because of their soft, nearly chitinless exoskeleton, silk moth larvae are readily digested by insectivorous predators, even those who only poorly chew their prey before swallowing.

Medical Disorders

There are several disease causing viruses, particularly *Reovirus* and *Fijivirus*, which have been isolated from infected silk moth larvae. One of the most severe often endemic diseases of these insects is cytoplasmic polyhedrosis virus infection. This RNA virus is transmitted from infected larvae to susceptible silkworms via the ingestion of infected material contaminating mulberry leaves. The virus attacks the midgut microvilli; very soon after they are infected, the larvae begin to lose condition from diminished absorption of nutrients from the intestine. There is no specific medical treatment for this condition. Several gram-negative bacteria and the protozoan *Nosema bombycis* have been isolated from these insects. If morbidity or multiple death losses are noted, the entire culture should be discarded and the container thoroughly cleansed before new larvae or adults are introduced.

Some wasps can employ the larvae of the silk moth as hosts upon which to lay their eggs. The developing wasp larvae feed upon the stung and immobilized silk moth larvae. Screening the culture container will effectively prevent this depredation.

Earthworms

The common earthworm, *Lumbricus vulgaris*, and its larger cousin, the "nightcrawler", *Lumbricus terrestris*, are easily grown in artificial culture. They require little care and their castings are a useful additive to house and garden plants.

Culture

Earthworms are most conveniently raised in well-drained wooden, plastic or *non-galvanized* metal trays or boxes filled with humus-enriched soil, leaf mold and/or planting mix which is kept slightly moist, but not soaking wet. The surface of the growing medium should be covered with moistened clean burlap sack cloth or corrugated cardboard slabs. The worms will tend to congregate beneath this cover, and are easily harvested as needed.

Nutrition

Earthworms ingest the soil-humus mix, absorbing what they can and passing the balance through their digestive systems as castings. All that is required is to add some cornmeal, clean salt-free table scraps, poultry mash, or well composted horse, cow or other animal manure to the growing medium occasionally. If deemed appropriate, the earthworm culture can be combined with a vegetable or flower garden box; such a symbiotic relationship benefits all concerned.

Reproduction

Earthworms are hermaphroditic and mate through most of the year, reaching a peak of activity during warm, moist weather conditions, particularly at night. Two worms stretch out from their burrows and bring their ventral surfaces together, with the anterior ends pointing in opposite directions. A specialized structure, the *clitellum*, on each worm grips five segments of the other worm. Specialized hair-like setae of each worm penetrate the body of the other to aid in holding the worms together. Each worm secretes a slime tube about itself, covering about 27 segment-like somites. On each worm, a pair of seminal grooves forms along which masses of sperm pass to enter the seminal recepticles of the other. The worms now separate from each other. Each worm later produces capsules or cocoons containing fertilized and embryonating eggs. These cocoons slip from the bodies of each mated worm, and in doing so, assume the shape of lemons as the central tube portion closes. The egg capsules are deposited in damp soil, and the young worms soon emerge and grow into adults.

Medical Disorders

Earthworms may be infected with *Bacillus thuringienis* and several yeasts and fungi. There are no specific treatments, although a few workers have tried feeding medicated feeds with mixed results.

True Crickets, Mormon "Crickets" and Similar Insects

The common house crickets, *Gryllus domesticus*, and *Acheta domestica*, and the Mormon cricket, *Anabrus simplex* (actually a wingless form of grasshopper) are cultured as food prey for insectivorous animals such as large spiders, scorpions, some fish, lizards, some birds, moles, shrews, bats, etc. Similarly, the Jerusalem crickets, *Stenopelmatus* also known as "sand crickets," Niños de la Tierra ("baby faces") and other forms of wingless grasshoppers, can be maintained in captivity as prey for insectivorous predators. The latter insects should be handled with forceps because they can deliver a nasty bite on the fingers of an unwary handler.

Culture

The common house crickets, *Gryllus domesticus* and *Acheta domestica*, are cultured in wooden or metal boxes or bins with tight-fitting, screened lids. Smaller numbers can be cultured in gallon jars, also fitted with screened screwcap tops, but adequate ventilation must be afforded to avoid excessive humidity.

Corrugated cardboard should be placed in the enclosure to serve as both hiding places and as nesting and egg deposition sites.

Nutrition and Water

Crickets are opportunistic feeders and are essentially omnivorous. They will eat a variety of vegetables, high protein cereal, cornmeal, poultry mash, small portions of dry kibbled dog food—and *each other*—if not supplied with sufficient amounts of available food or moisture. They obtain most of this moisture from what they eat, but also will drink water from shallow containers such as jar lids or bottle caps in which small pieces of cellulose sponge are fitted. The insects will imbibe moisture from the wet sponge and will be prevented from being trapped and drowned.

Reproduction

Fertilization is internal and individual eggs are laid in and on the cardboard substrate. The baby crickets, called "pinheads," may be fed, as needed, to whatever they were bred to provide food for or allowed to increase in size and numbers to augment the breeding population.

Mealworms

The larvae of the grain beetle, *Tenebrio molitor*, have been raised for many years as fish bait and for feeding fish, amphibians, reptiles, birds and small insectivorous mammals. Usually it is the larvae that are employed as food items. If allowed to develop to maturity, the larvae will eventually undergo metamorphosis, becoming first pupae and then mature adult beetles. When necessary, one can select those larvae that have recently undergone periodic molting (ecdysis). Immediately after shedding their outer covering, the larvae are paler in color and lack the dark brown chitinous body rings. Within about 24 hours the chitinous exoskeleton hardens and darkens. The advantage of feeding these soft larvae is that they are more easily chewed and thus will yield more of their potential energy than when they are covered with a harder exoskeleton.

Culture

Mealworms can be grown in metal, plastic or glass containers that are flat and broad and that are fitted with screened covers that will prevent the escape of beetles and larvae. The screens also prevent the contamination of the growing medium or predation by moths and other extraneous competitive insects. The surface of the growth medium should be covered with clean burlap sackcloth; this will serve as a hiding place and also as a brooding area. Note that some mealworms that are sold commercially for fish bait have been treated with juvenile hormone which, although causing these larvae to grow larger, will also inhibit their completion of metamorphosis.

Nutrition

In order to produce mealworms that are nutritionally sound, especially with respect to calcium, the growing medium of wheat bran midlings should be supplemented with 15% alfalfa flakes, 20% high protein baby cereal, and 15% dry poultry mash. Moisture is provided with freshly halved apples and potatoes that are replaced as they are consumed or become moldy. Excess larvae can be stored in clean dry bran under refrigeration for several weeks. Refrigeration will help prevent metamorphosis from occurring.

Several chemical analyses have been performed upon *Tenebrio molitor* larvae. Two are reproduced in Table 11.

Note that the ratio of calcium to phosphorus is grossly out of balance with respect to that required for the normal growth and maintenance of bone matrix. Mealworms remain an inade-

Table 11
Chemical Composition of Meal Worms

	a	b
Fat	12.00%	9.37%
Protein	20.80	20.75
Carbohydrate	1.86	
Fiber	2.00	
Moisture	61.90	67.41
Calcium	0.03	
Phosphorus	0.27	
Ash		4.70
Calories (per 100 grams)	204.00	

Sources: *a*: Roots, C.: The Chemical Composition of Mealworms. *WRC Wildlife J.* 1(4), 1978; *b*: Hills Division of Riviana Foods, Inc. Also see Redford and Dorea, 1984.

quate sole diet for insectivorous reptiles, although supplementing the culture medium in which they are grown will improve their nutritive value.

Reproduction

Sexually mature meal beetles mate and deposit their fertile eggs in or on the burlap sacking material. As the young larvae hatch out, they burrow into the rich growing medium and should be harvested as they are required to meet the demands of the collection.

Diseases

Mealworms and adult *Tenebrio* beetles are susceptible to several bacterial, mycotic, and viral pathogens: *Bacillus thuringiensis*, *Serratia marcescens*, *Monilia anisopliae*, *Baculovirus* and assorted beta endotoxins have been isolated from clinically abnormal meal-worm larvae and adult beetles.

Fruit Fly and House Fly Culture

To some readers, it might seem strange to go purposely out of one's way to produce creatures that are common household pests. As anyone who has ever shopped in the produce section of a supermarket knows, fruit flies are frequently present—particularly if ripe pineapples or bananas are on display. An hour or two after you bring these fruits home, these tiny insects may be found swarming around your fruit bowl. For students of biology or genetics, the fruit fly, *Drosophila melanogaster*, has been one of the primary animal models for the study of Mendelian inheritance. The successful culture of newly hatched or newly born tiny insectivorous lizards, praying mantids, spiders, etc. demands a steady supply of small soft-bodied insects. The discovery and propagation of genetically wingless forms of fruit flies now make it possible to produce any number of flightless insects that can only crawl about. The culture medium upon which these flies grow adds necessary nutrients to the insectivores' diets also.

If larger flies are required, the common house fly, *Musca domestica*, can be grown. These flies are cultured on a formula that is a modification of one that has been employed at the Department of Genetics at the University of California, Davis. This modification contains meat extract and was described by Morris (1983) (Table 12).

Morris stated that the dextrose/sucrose mixture could be replaced by substituting 125 ml molasses or 125 ml of a 50:50 molasses/liquid dextrose (or Karo syrup) mix.

Mix the dry ingredients and store in a cool place until needed to make a fresh batch of growing medium. When required, add warm water to dissolve the sugar/syrup mix so that the resulting solution does not contain lumps. Then add the agar, brewer's yeast, and cornmeal. Meanwhile boil the water. Pour the diluted sugar/syrup/agar/yeast/cornmeal mix into the boiling water, stir well, and allow to cool. After it becomes sufficiently cool to handle comfortably, pour the mixture into 50 ml glass vials, taking particular care to avoid splashing the culture medium onto the sides of the vials. The recipe above is sufficient to produce approximately 200 vials. Place a piece of tissue paper or lens tissue into the medium to afford the fly larvae a place upon which to pupate. The open end of the vials should be closed with a cotton ball wrapped in a piece of cotton gauze or nylon stocking.

Various strains of fruit flies can be purchased from any of several biological supply firms. If only a few flies are needed, a call to the genetics department of most local colleges or univer-

Table 12
Culture Medium Formula

agar	18 gm
brewer's yeast	28 ml
cornmeal	56 ml
dextrose	126 ml
sucrose	65 gm
warm water	500 ml
boiling water	1,000 ml
calcium proprionate*	9 ml

*Catalog number 1053347 Eastman Kodak Co., Inc.

sities might result in a starter culture. As mentioned previously, the genetically wingless or vestigial winged (vg) strains are preferable to those that can fly.

Breeding occurs in the vials, so to "inoculate" an uncultivated vial, pour a few flies from a previously thriving vial into each freshly prepared vial. The generation time of these insects is brief; to produce a steady supply of flies, plan to inoculate a new set of vials about every 5–7 days. The number of vials to be incubated will depend upon the demand for flies. Actively crawling flies can be slowed or immobilized by placing the vial(s) in the refrigerator for a few minutes. Larger house flies can be raised on artificial culture media much like that described above, but some form of high quality animal protein such as meat extract should be added. Of course, a much larger culture container must be employed. The old-fashioned glass milk or cream bottles with large-diameter tops are excellent for this purpose. The tops of these containers should be stoppered with gauze-covered cotton that will confine the flies, yet permit adequate air circulation.

Although the culture of fruit flies is not necessarily accompanied by an unpleasant odor, the production of housefly maggots can be associated with a foul smell because the maggots themselves and their wastes have a disagreeable odor.

Wax Moth Larvae Culture

The wax moth, *Galleria mellonella*, and the lesser bee moth, *Achroia grisella*, are predators of honey bees' brood combs. Frequently, these combs are deserted, but if wax moths attempt to enter an occupied beehive, a healthy colony of bees usually is able to combat and repel the depredations of these insects. The larvae of these moths have been cultured as prey for insectivorous invertebrates and vertebrates.

A modified culture technique has been employed by the Department of Nematology at the University of California, Davis. It was described by Morris (1983) as is printed here with further modifications. The diet is altered as the size of the larvae increases (Table 13). Table 14 lists sources of living and frozen reptile food.

Table 13
Culture Technique for Wax Moth Larvae

Culture Medium for Newly Hatched Larvae

boiling water	100 ml
glycerine, U.S.P.	100 ml
honey	100 ml
vitamin supplement (Vionate, or Avitron)	5 gm/5 ml
infant cereal	1,200 gm
calcium proprionate*	1.5 ml

Bake the cereal at 200°F for 2 hours. Mix the liquid ingredients together. Add vitamin supplement and calcium proprionate, mixing well. Add liquids to cereal, mixing thoroughly to distribute all the ingredients. Store in glass jars. This growth medium is used to feed moth larvae until they can be transferred to a less refined diet.

Diet for Larger Larvae

miller's bran	1,200 ml (dry measure)
honey	200 ml (dry measure)
water	100 ml (dry measure)
glycerine, U.S.P.	9 ml (dry measure)
brewer's yeast	1 tablet
vitamin supplement (Vionate, or Avitron)	5 gm/5 ml
calcium proprionate*	1.5 gm

Bake the miller's bran at 200°F for 2 hours. Thoroughly mix dry ingredients with liquid ingredients until homogeneous. Place 2–4 tablespoons of the medium into one-gallon, wide-mouth glass pickle, mayonnaise, or mustard jars. Place half-grown (¼" or larger) larvae into jar, and cover the top of the container with one or more layers of cotton gauze or nylon stocking material. Depending upon the size of the culture and jar and the number of larvae (or adult moths) that are required, 2–10 pairs are used to start a culture.

Place culture jars in an environmental temperature of about 27.7° to 34°C (82–94°F). The larvae will mature and pupate within approximately two weeks, and the moths will soon emerge to mate and produce another generation.

Source: Morris (1983).
*Catalog number 1053347 Eastman Kodak Co., Inc.

The sexes are easy to determine because the male moths are smaller. A piece of accordion-pleated filter or wax paper placed in the jar will mimic a brood comb and afford the moths a place upon which to deposit their fertilized eggs. If the demand is sufficiently large, the culture jars should be alternated so that a continual supply of larvae and moths will be available.

Table 14
Some Sources of Living and Frozen Reptile Food

Mealworms, Wax Moth Larvae, Fly Larvae, Silk Moth Eggs

Carolina Biological Supply Co.
Eastern United States
  Burlington, NC 27215
  1-919-584-0381
  1-800-547-1733
Western United States
  Powell Laboratories
  Gladstone, OR 97027
  1-503-656-1641
  1-800-334-5551

Grubco, Inc.
P.O. Box 15001
Hamilton, OH 45015
1-800-222-3563

Niles Biological Supply
9298 Elder Creek Rd.
P.O. Box 191543
Sacramento, CA 95829
1-916-739-8048

Rainbow Mealworms
P.O. Box 4525
126 E. Spruce St.
Compton, CA 90220

Delectables
%ARBICO
P.O. Box 4247 CRB
Tucson, AZ 85738
1-800-SOS-BUGS

Crickets

Armstrong's Cricket Farm
P.O. Box 125
W. Monroe, LA 71294-0125
1-318-387-6000

Live and Frozen Mice and Rats

Karen Ellett
Box 78, Rt #1
Greenville, NY 12083
1-518-966-8192

Kelly Haller
4236 S.E. 25th
Topeka, KS 66605
1-913-234-3358

Glades Herpetoculture
P.O. Box 643
Alva, FL 33920
1-813-675-6423

The Gourmet Rodent, Steve Hammond
%Exceptional Exotics
12319 Steedland
Louisville, KY 40229
1-502-955-8705

William Bryant
1123 SR 346A
Archer, FL 32618
1-904-495-9024

Michael J. Miller, D.V.M.
10504 S. Roberts Road
Palos Hills, IL 60465
1-312-974-2600

Charles Hicks
705 River Road, #102
San Marcos, TX 78666
1-512-754-0056

Marv's Small Animals
138A Annie Glidden Road
Kingston, IL 60145
1-815-784-3461

RANPRO
Box 421115
Del Rio, TX 78842-1115
1-512-774-4353

Eric Richter
1-708-863-7418

SPECIAL RECIPES FOR AQUATIC TURTLES

Some workers in the field of reptilian nutrition have developed formulae for feeding chelonians, particularly aquatic and semi-aquatic species (Czajka, 1981; Feldman, 1980; Frye, 1973, 1974, 1981, 1986). These preparations rely upon unflavored gelatin or agar-agar as binding agents and a variety of either fresh meats or commercial dog, cat, monkey, trout, catfish, or shrimp chows or high protein infant cereal, as sources of balanced protein, vitamins, and minerals. Frye (1981, 1986) recommended the addition of pelleted ground alfalfa rabbit or guinea pig chow as a source of natural carotene and roughage. An additional source of calcium such as calcium carbonate or ground hen's egg shells is also recommended. The dry materials are first thoroughly mixed and then overlayed and well mixed with warm agar-agar or unflavored gelatin and allowed to cool in shallow pans. Once completely cooled and firm, the resulting thick sheets of turtle feed are cut into blocks and frozen in air-tight plastic bags or freezer containers for later thawing and use. These recipes have the advantage of keeping the particulate foodstuffs from dissolving too quickly and fouling the tank water. The exact recipe is not as important as the inclusion of a readily available source of high quality animal proteins, nutritive roughage, and vitamin-mineral complement. The palatability must, of course, be excellent because even the most nutritious formula for a captive diet will be useless if it is not readily eaten. Lastly, great attention must be paid to avoid either under- or oversupplementation of lipid-soluble vitamins and some minerals. This warning applies particularly to vitamins A and D. Carotene is the natural green-yellow-orange plant pigment from which the animals synthesize vitamin A; unlike vitamin A and D, carotene is essentially not toxic. Vitamin D can be synthesized by the animals when they are exposed to unfiltered ultraviolet light of appropriate wavelength.

Today, there are a few excellent commercial products available at most pet and aquarium dealers that have nearly obviated the necessity for making home-concocted formulae, except for a few unusual instances. Some of these products are mentioned in the next section.

Eugenia, hibiscus, nasturtium and wandering Jew are easily grown in small containers and make colorful additions to cage enclosures.

Small fish are easily raised for food items required in diet(s) of a captive reptile collection. Wild-caught sticklebacks and mosquito fish should be avoided because they are the natural vectors for several serious parasites of fish-eating reptiles.

Commercial Products Suitable for Turtlefood

Several readily available commercial feeds have shown great promise as dietary staples for aquatic turtles and terrapins. Of these, ReptoMin (TetraWerke) was the only one specifically formulated for reptiles, but the others appear to be well accepted. These products and their manufacturers are listed in Table 15.

HOW TO PREPARE, STORE, AND USE FROZEN FOOD

Special food items for reptiles with highly limited dietary requirements may not be available year-round. However, many, if not most, feeds can be kept in frozen storage if they are properly prepared for freezing and carefully handled during thawing and when presented to the animals.

Some feeds have a substantially shorter frozen storage life than others. The storage life generally depends upon not only the nature of the feed, but also the manner in which it was prepared prior to and after freezing. For example, if fish are "flash" frozen immediately after they are caught, little postmortem decomposition occurs. However, if the fish are permitted to remain at higher than freezing temperature for several hours before they are frozen, their tissues will contain lesser amounts of vital nutrients and their storage longevity will be significantly reduced.

Many plants and animals contain endogenous tissue enzymes that reduce the inherent nutrient quality that can be gained when the food items are fed. Immersing them briefly in

Table 15
Commercial Products Suitable for Turtle Food

Product and Manufacturer	Comment
Tetra ReptoMin Floating Food Sticks TetraWerke Federal Republic of Germany Available at most pet and aquarium dealers	Worm-like shape, floats on surface; relatively expensive; highly palatable
Tetra TabiMin TetraWerke Available at most pet and aquarium dealers	Tablets, sink to bottom, slowly dissolve; relatively expensive
Trout Chow Ralston-Purina St. Louis, MO Available at most pet and aquarium dealers	Semi-round pellets, floats; inexpensive
Catfish Chow Ralston-Purina Available at most pet and aquarium dealers	Semi-round pellets, floating or sinking; inexpensive
Shrimp Chow Ralston-Purina Available at most pet and aquarium dealers	Small pellets, floating or sinking; inexpensive
Tender Vittles Cat Food Ralston-Purina Available at most grocery stores	Semi-moist short strands, sinks; high level of vitamins A & D; inexpensive

boiling water inactivates the enzymes that cause spoilage and shorten storage time. Generally, a period of 15–30 seconds is sufficient to cause this enzymatic inactivation.

Some water is lost from the frozen food. This generally slow dehydration, termed "freezer burn," can be diminished by vacuum packaging or wrapping carefully to exclude as much air as possible. Tough polyethylene plastic self-sealing (Ziploc) bags are particularly useful for storing food items in a freezer. Another method is to employ plastic packaging machines that vacuum pack and then heat seal the food prior to freezing.

Properly fast-frozen and stored mice, rats, and chicks will retain their freshness and nutrient quality for up to 6 months. In order to reduce the likelihood of freezer burn, the frozen prey animals should be stored in bags containing only enough for a week or two; after each portion is removed, excess air must be expelled before the package is resealed.

Similarly, commercially frozen mixed vegetables should be transferred from the boxes or nonresealable bags in which they are sold to resealable plastic freezer storage bags so that excess air can be excluded before the bag is placed back in the freezer. Unopened packages of commercially frozen vegetables can be kept for 6–12 months, but maximum nutrient value is obtained during the first 3–6 months. Properly frozen insects are best fed within 3 months.

The period from thawing to feeding should be as short as possible. One method for thawing small mammals and birds is to use a microwave oven set to operate at its thaw-cycle for a brief time. An alternative method uses immersion in tepid water. If water is used to thaw frozen animals for snake food, the food items should be temporarily sealed in a dry storage bag so that they do not become water-logged. Frozen vegetables need only be placed at room temperature for a brief time before they are fed to herbivorous animals.

Appendix

Directory of Herpetological Societies of the United States

Arizona

Arizona Herpetological Association
P.O. Box 39127
Phoenix, AZ 85069-9127

National Turtle & Tortoise Society
P.O. Box 9806
Phoenix, AZ 85068-9806

Southern Arizona Herpetological Association
%Tom Boyden
4521 West Mars Street
Tucson, AZ 85704

Tucson Herpetological Society
P.O. Box 31531
Tucson, AZ 85751-1531

Arkansas

Arkansas Herpetological Society
%Floyd Perk
Route 2, Box 16
Hensley, AK 72065

California

American Federation of Herpetoculturists
P.O. Box 1131
Lakeside, CA 92040

Bay Area Amphibian & Reptile Society
Palo Alto Junior Museum
1451 Middlefield Road
Palo Alto, CA 94301

California Turtle & Tortoise Society
P.O. Box 8952
Fountain Valley, CA 92728

Island Empire Herpetological Society
San Bernardino County Museum
2024 Orange Tree Lane
Redlands, CA 92373

Northern California Herpetological Society
Box 1363
Davis, Ca 95617-1363

Sacramento Valley Herpetological Society
%Bob Pedder
6007 Watt Avenue
North Highlands, CA 95660

San Diego Herpetological Society
P.O. Box 4439
San Diego, CA 92104-4439

San Diego Turtle & Tortoise Society
%13963 Lyons Valley Road
Jamul, CA 92035-9607

Southern California Snake Association
P.O. Box 2932
Santa Fe Springs, CA 90670

Southwestern Herpetological Society
P.O. Box 7469
Van Nuys, CA 91409

Colorado

Colorado Herpetological Society
P.O. Box 15381
Denver, CO 30215

Northeast Colorado Herpetological Society
%Roger Klingenberg
6247 West 10th
Greeley, CO 80631

Rocky Mountain Herpetological Society
%Dave Baker
605 W. Colorado Avenue
Colorado Springs, CO 80905

Connecticut

Connecticut Herpetological Society
%George Whitney, D.V.M.
860 Oakwood Road
Orange, CT 06477

Eastern Seaboard Herpetological League
%Michael Uricheck
77 Faber Avenue
Waterbury, CT 06704

Delaware

Delaware Herpetological Society
Ashland Nature Center
Brackenville & Barley Mill Road
Hockessin, DE 19707

District of Columbia

American Society of Icthyologists & Herpetologists
National Museum of Natural History
Washington, DC 20560

Washington Herpetological Society
%Frank Watrous III
12420 Rock Ridge Road
Herndon, VA 22070

Florida

Central Florida Herpetological Society
P.O. Box 3277
Winter Haven, FL 33881

Florida Panhandle Herpetological Society
%The Zoo
5801 Gulf Breeze Parkway
Gulf Breeze, FL 32561

Florida West Coast Herpetological Society
%John Lewis
1312 S. Evergreen Avenue
Clearwater, FL 33516

Gainesville Herpetological Society
P.O. Box 7104
Gainesville, FL 32605

Gopher Tortoise Council
%Patricia Ashton
611 NW 79th Drive
Gainesville, FL 32607

Herpetological League
Department of Biology
University of Miami
Coral Gables, FL 33124

Palm Beach County Herpetological Society
%Greg Longhurst
P.O. Box 125
Lozahatchee, FL 33470

Tampa Bay Herpetological Society
%3310-A Carlton Arms Drive
Tampa, FL 33614

Georgia

Georgia Herpetological Society
Reptile House, Zoo Atlanta
800 Cherokee Avenue, SE
Atlanta, GA 30315

Troup County Association of Herpetologists
%C.W. Dodgen
801 Grant Street
LaGrange, GA 30240

Idaho

Idaho Herpetological Society
P.O. Box 6329
Boise, ID 83707

Illinois

Central Illinois Herpetological Society
1125 West Lake
Peoria, IL 61603

Chicago Herpetological Society
2001 N. Clark Street
Chicago, IL 60614

Indiana

Hoosier Herpetological Society
%Dennis Brown
2906 South Taft
Indianapolis, IN 46241

Mid-Mississippi Valley Herpetological Society
%Mike Ladoto
925 Park Place Drive
Evansville, IN 44714

Iowa

Iowa Herpetological Society
%P.O. Box 23035
Des Moines, IA 50322

Kansas

Kansas Herpetological Society
Museum of Natural History
University of Kansas
Lawrence, KS 66045

Kaw Valley Herpetological Society
Route 1, Box 29B
Eudora, KS 66025

Louisiana

Louisiana Herpetological Society
Museum of Natural Sciences
Foster Hall
Louisiana State University
Baton Rouge, LA 70803

Maryland

Maryland Herpetological Society
Natural History Society of Maryland
2643 N. Charles Street
Baltimore, MD 21218

Massachusetts

Massachusetts Herpetological Society
P.O. Box 1082
Boston, MA 02103

Western Massachusetts Herpetological Society
Science Museum
236 State Street
Springfield, MA 01103

Michigan

Great Lakes Herpetological Society
%Jeff Gee
4308 North Woodward
Royal Oak, MI 48072

Michigan Society of Herpetologists
%321 West Oakland
Lansing, MI 48906

Minnesota

Minnesota Herpetological Society
Bell Museum of Natural History
10 Church Street, SE
Minneapolis, MN 55455

Mississippi

Southern Mississippi Herpetological Association
P.O. Box 10047
Gulfport, MS 39505

Missouri

St. Louis Herpetological Society
P.O. Box 9216
St. Louis, MO 63117

Nebraska

Nebraska Herpetological Society
Biology Department
University of Nebraska-Omaha
Omaha, NE 68182

Nevada

Northern Nevada Herpetological Society
%Bill Gill
348½ Wheeler Avenue
Reno, NV 89502

New Mexico

New Mexico Herpetological Society
Department of Biology
University of New Mexico
Albuquerque, NM 87131

New York

Long Island Herpetological Society
117 E. Santa Barbara Road
Lindenhurst, NY 11757

New York Herpetological Society
Box 1245, Grand Central Station
New York, NY 10017

New York Turtle & Tortoise Society
163 Amsterdam Ave., Suite 365
New York, NY 10023

North Carolina

North Carolina Herpetological Society
State Museum of Natural Sciences
P.O. Box 27647
Raleigh, NC 27611

Ohio

Greater Cincinnati Herpetological Society
%Museum of Natural History
1720 Gilbert Avenue
Cincinnati, OH 45202

Greater Dayton Herpetological Society
%Museum of Natural History
2629 Ridge Avenue
Dayton, OH 45414

Northern Ohio Association of Herpetologists
Department of Biology
Case Western Reserve University
Cleveland, OH 44106

Society for the Study of Amphibians and Reptiles
Department of Zoology
Miami University
Oxford, OH 45056

Oklahoma

Oklahoma Herpetological Society
%Patrick Mulvany
7315 East 81st Place
Tulsa, OK 73133

Oregon

Oregon Herpetological Society
%Steven Aveldson
8435 Derbyshire Lane
Eugene, OR 97405

Pennsylvania

Lehigh Valley Herpetological Society
%G. Leonard Knapp
215 Lawn Avenue
Sellersville, PA 18960

Philadelphia Herpetological Society
%Mark Miller
9573 Walley Avenue
Philadelphia, PA 19115-3009

Susquehanna Herpetological Society
%Sam Burleigh
211 S. Market Street
Muncy, PA 17756

Texas

Dallas Herpetological Society
P.O. Box 3672
Irving, TX 75015

East Texas Herpetological Society
P.O. Box 1561
Trinity, TX 75862

Greater San Antonio Herpetological Society
%W. Rowe Elliott
134 Aldrich Street
San Antonio, TX 78227

Lubbock Turtle & Tortoise Society
%Joe Cain
5708 64th Street
Lubbock, TX 79424

North Texas Herpetological Society
P.O. Box 470771
Fort Worth, TX 76147

South Texas Turtle & Tortoise Society
%James Maples, Jr.
927 Wilson
Alice, TX 78332

Texas Herpetological Society
HC 53, Box 3225
Bulverde, TX 78163

Utah

Utah Herpetological Society
Hogle Zoological Gardens
P.O. Box 8475
Salt Lake City, UT 84108

Virginia

Virginia Herpetological Society
%Joseph Mitchell
Dept. of Biology
University of Richmond
Richmond, VA 23173

Washington

Pacific Northwest Herpetological Society
1308 North 8th Street
Tacoma, WA 98403

Wisconsin

International Society for the Study of Dendrobatid Frogs
P.O. Box 366
Germantown, WI 53022

Wisconsin Herpetological Society
9137 W. Mill Road
Milwaukee, WI 53225-1701

*Compiled by the Reptile and Amphibian Magazine and *Herpetological Review*

Appendix

Glossary of Terms

ambient: environmental, surrounding (temperature).

anomaly: an abnormal condition or conformation involving an organ.

anorexia: loss of appetite; refusal to feed.

antemortem: before death.

appendicular skeleton: pertaining to the limbs and their appendages.

aquatic: water living.

arboreal: tree-living or climbing.

atrophy: wasting of tissue or an organ.

axial skeleton: pertaining to skull, vertebral column and pelvis.

bioactive: having an effect on living tissues in an organism.

calcareous: chalky, calcium-rich.

cancellous (trabecular) bone: spongy bone; bony tissue characterized by its spaces, often filled with bone marrow.

carapace: the upper shell of a tortoise, turtle or some invertebrates.

carnivorous: meat or flesh-eating; confining one's diet to meat.

carrion: the remains of a dead animal.

caudal: pertaining to the tail end.

chelonian: a turtle, tortoise or terrapin.

chitin: a horny, insoluble polysaccharide from which the exoskeleton of many invertebrates is formed.

cloaca: the common vault into which the alimentary and genitourinary systems terminate via their own tubes or ducts.

conspecific: sharing the same species.

constipation: difficulty or inability to pass feces.

copradeum: the portion of the terminal digestive tract that empties into the cloaca.

cortical bone: the dense and outermost portion of bone.

cranial: pertaining to the head end of an animal.

coumadin: a drug used to delay blood clotting.

cytology: the microscopic study of cells making up a tissue or organ.

diapause: a resting, biologically inactive phase; in this usage, pertaining to insect egg development.

dorsal: pertaining to the back.

dyspnea: difficult or labored breathing.

ecdysis: the process of molting or shedding of the epidermis.

ectothermic: having a body temperature related to the environmental temperature, rather than being able to produce heat internally.

electrolytes: physiologic chemical compounds contained in tissue or body fluids.

endogenous: originating inside an organism; produced by an internal organ or system.

endoscopy: the process of examining the internal structures of an organism by use of a light-equipped medical instrument.

etiology: the cause for some condition.

exogenous: originating outside an organism.

exoskeleton: the non-bony outer covering of many invertebrates, often composed of insoluble chitin.

flora: referring to plants.

gavage: washing or placing a substance into a hollow organ via a tube or other device.

geophagy: earth-eating.

gingival: referring to the gums.

granuloma: an inflammatory lesion containing an infectious, parasitic, or foreign object; usually surrounded by dense connective tissue; often follows an abscess.

halogen: a sodium salt containing bromine, chlorine, fluorine, or iodine. Some plants can concentrate these compounds from the soil in which they grow.

herbivorous: plant- or vegetable-eating.

hydroponic: plant growing in a water-nutrient solution rather than in soil.

hydroxyapatite: a calcium-containing salt from which bone is formed; occasionally, hydroxyapatite is deposited in abnormal tissues.

hyperparathyroid: overactive secretion by the parathyroid glands.

hyperplasia: abnormal cell number.

hypertonic: a solution which has a greater than a physiologic or isotonic concentration.

hypertrophy: abnormal cell or organ size.

hypervitaminosis: pertaining to a specific vitamin overdosage.

hypotonic: a solution which has a less than physiologic or isotonic concentration.

hypovitaminosis: pertaining to a specific vitamin deficiency.

iatrogenic: a condition caused by a treatment.

idiopathic: of unknown cause or etiology.

ileus: obstruction of the intestines; can be caused by a physical object or by a failure of normal peristaltic movement.

inanition: partial or complete starvation resulting from a lack of food.

indigenous: native.

ingesta: swallowed food.

insectivorous: insect-eating; confining one's diet to insects.

inspissated: abnormally dry.

integument: skin.

invertebrate: lacking a backbone or vertebral column.

intercurrent: a condition occurring at the same time as another.

isotonic: having the same concentration; with respect to physiologic solutions that are compatible with living tissues.

keratin: a constituent of epidermis and integumentary shell plates.

lateral: pertaining to the side or flank.

lithophagy: stone- or rock-eating.

maceration: softening of a solid by the action of water or other aqueous solution.

mandible: the lower jaw.

mastication: chewing.

metazoa: multicellular animals.

metabolites: constitutents of food or metabolism used or excreted by an organism.

microflora: bacterial and protozoan inhabitants of the intestine.

motility: ability to move; used here in the context of moving gastrointestinal contents through the alimentary tract.

necropsy: examination of a corpse.

neoplasm: new-growth; a tumor.

nephrotoxic: chemically damaging to the kidneys.

neuropathy: a disorder of a nervous organ or tissue.

obstipation: partial or complete obstruction to the passage of feces through the intestine.

omnivorous: eating both animal flesh and plant material.

ophiophagous: snake-eating; confining the diet to snakes.

osteomalacia: softening of the bones.

osteopenia: insufficiently mineralized bone.

oviphagous: egg-eating.

parathyroid: one or more pairs of glandular structures whose hormone secretion, parathormone, is responsible for calcium and phosphorus metabolism.

pathogen: a disease-causing organism.

pelagic: marine; sea-going.

physiologic (fluids): fluids mimicking those that are present in an organism; compatible with the natural-occurring fluids characteristic of an organism.

piscivorous: fish-eating.

plasma: the fluid portion of unclotted blood.

plastron: the lower shell portion of a tortoise, turtle or terrapin.

postmortem: after death.

proctodeum: the terminus of the alimentary tract.

puree: finely chopped or strained food.

pyogranuloma: a chronic inflammatory pus-filled enlargement.

radiography: diagnostic X-ray imaging.

renal: pertaining to the kidneys.

rostral: pertaining to the beak or nose of an animal.

sclerotized: hard, tough, impervious.

sequester: to take up and concentrate a compound or chemical from the environment and store in the tissues; used here in the context of a plant or animal concentrating selectively storing soil elements in the form of their sodium salts.

slurry: a soft, finely ground, near-liquid suspension of a material, in an aqueous agent.

spongivorous: sponge-eating.

squamous: flattened.

squamate: referring to possessing scales; snakes and lizards are squamate reptiles.

subcutaneous: referring to beneath the skin.

taxa: a group of animals sharing similar characteristics common to their species.

taxonomy: the naming of or identification of a group of organisms or objects.

terrestrial: land-living.

tympany: bloating; the retention of gastrointestinal gas.

urates: urinary salts of uric acid, usually sodium, potassium or ammonium.

urodeum: that portion of the terminus of the urinary excretory system that empties into the common cloaca.

ventral: pertaining to the belly .

vestigial: poorly developed; a remnant form of an organ.

xeroderma: pathologically dry, thickened, and scaly skin.

Appendix C
References

Allen, M.E. Dietary Induction and Prevention of Osteodystrophy in an Insectivorous Reptile *Eublepharis macularius*: Characterization by Radiography and Histopathology. *Proc. III Int. Coll. Pathol. Reptiles Amphibians.* Orlando, FL, January 13–15, 1989.

―――― and Capen, C.C. Fine Structural Changes of Bone Cells in Experimental Nutritional Osteodystrophy of Green Iguanas. *Virch. Arch.*, B. 20:169–184; 1976.

――――.Ultrastructural Evaluation of Parathyroid and Ultimobranchial Glands in Iguanas with Experimental Nutritional Osteodystrophy. *Gen. Comp. Endocrinol.*, 30(2):209–222; 1976.

――――, Oftedal, O.Y., and Knapka, J. Manipulation of Calcium and Phosphorus Levels in Live Prey. In: *Proc. of NE Section on Amer. Assoc. Zoos, Parks, and Aquar.* 1982.

Appleby, E.C. and Siller, W.G. Some Cases of Gout in Reptiles. *J. Path. Bact.*, 80:427–430; 1967.

Auffenberg, W. Feeding Strategy of the Caicos Ground Iguana, *Cyclura carinata*. In: *Iguanas of The World: Their Behavior, Ecology and Conservation*. ed. G.M. Burghardt and A.S. Rand, 84–116. Park Ridge, NJ: Noyes Publishing. 1982.

――――. *Gray's Monitor Lizard*. Gainesville, FL: Univ. Florida Press, 1988.

Banks, C.B. *Naja melanoleuca*: Copraphagy. *Herp. Rev.*, 15(4): 113; 1984.

――――. Reproductive History of a Captive Colony of *Iguana iguana*. *Acta Zoo. et Path. Antverpiensia*, 78: 101–114; 1984.

Barten, S.L. The Induction of Voluntary Feeding in Captive Snakes. *Bull. Chi. Herp. Soc.*, 16(1):1–5; 1981.

――――. Herp Health & Husbandry Hints: Calcium Metabolism When Feeding Herps Pinky Mice. *Bull. Chi. Herp. Soc.*, 23(1):14–15,1988.

Bartlett, R.D. Notes on Some Insular and "Dwarf" Boas of the New World. *Reptile & Amphibian Mag.*, May/June, 1990.

Bauer, A.M. Extracranial Endolymphatic Sacs in *Eurydactylodes* (Reptilia: Gekkonidae), with Comments on Endolymphatic Function in Lizards. *J. Herp.*, 23(2):172–175; 1989.

Belkin, D.A. Reduction of Metabolic Rate in Response to Starvation in the Turtle, *Sternotherus minor. Copeia*, 1965(3):367–368; 1965.

Blomhoff, R., Green, M.H., Berg, T., and Norum, K.R. Transport and Storage of Vitamin A. *Science*, 250:239-240; 1990

Bradshaw, S.D. and Shoemaker, V.H. Aspects of Water and Electrolyte Changes in a Field Population of Amphibolurus Lizards. *Comp. Biochem. Physiol.*, 20:855–865; 1967.

Brambel, C.E. Prothrombin Activity of Turtle Blood and the Effect of a Synthetic Vitamin K Derivative. *J. Cell Comp. Physiol.*, 18:221–232; 1941.

Britton, S.W. and Klein, R.F. Emotional Hyperglycemia and Hyperthermia in Tropical Mammals and Reptiles. *Am. J. Physiol.*, 125:730–734; 1939.

Bustard, H.R. Reproduction in the Australian Gekkonid Genus *Oedura* Gray, 1842. *Herpetologica*, 23:276–284; 1967.

─────. The Egg-Shell of Gekkonid Lizards: a Taxonomic Adjunct. *Copeia*, 1968:162–164; 1968.

───── and Maderson, P.F. The Eating of Shed Material in Squamate Reptiles. *Herpetologica*, 21:306–308; 1965.

Capen, C.C. and Marten, S.L. Calcium-Regulating Hormones and Diseases of the Parathyroid Glands. In: *Textbook of Veterinary Internal Medicine*, 2nd Ed., ed. S.J. Ettinger, 1561–1565. Philadelphia: W.B. Saunders, 1983.

Chatterjee, G.C. Effects of Ascorbic Acid Deficiency in Animals. In: *The Vitamins*, 2nd Ed. Vol. 1. ed. Sebrell, W.H., Jr. and R.E. Harris. 407–457. New York: Academic Press, 1967.

Chatterjee, I.B. Biosynthesis of L-Ascorbate in Animals. In: *Methods in Enzymology*, Vol. XVIII, Part A. eds. McCormic, D.B. and L.D. Wright, 28–34. New York: Academic Press, 1970.

─────. Vitamin C. Synthesis in Animals: Evolutionary Trend. *Science and Culture*, 39:210–212; 1973.

Collins, D. Quantities of Calcium Carbonate Needed to Balance Calcium-Phosphorus Ratios of Various Meats. *J. Zoo Anim. Med.*, 2: 25; 1971.

Coulson, R.A. and Hernandez, T. Glucose Studies in Crocodilia. *Endocrinology*, 53:311–320; 1953.

─────. Renal Failure in the Alligator. *Am. J. Physiol.*, 200:893–897; 1961.

─────. *Biochemistry of the Alligator. A Study of Metabolism in Slow Motion.* Baton Rouge: Louisiana State University Press, 1964.

Czajka, A.F. Gelatin-Bonded Food for Turtles. *Bull. Chicago Herp. Soc.*, 16(2):40–41; 1981.

Dacke, C.G. *Calcium Regulation in Sub-Mammalian Vertebrates.* New York: Academic Press. 1979.

Demovsky, R. and Greenberg, L.D. Growth Effect and Tissue Distribution of Vitamin A Following Intravenous Injection of Vitamin A-rich Chyle. *Proc. Soc. Exptl. Biol. Med.*, 118:158–161; 1964.

Dantzler, W. H. Effect of Metabolic Alkalosis and Acidosis on Tubular Urate Secretion in Water Snakes. *Am. J. Physiol.*, 215 (3):747–751; 1968.

───── and Schmidt-Nielsen, B. Excretion in Fresh-Water Turtle (*Pseudemys scripta*) and Desert Tortoise (*Gopherus agassizi*). *A. J. Physiol.*, 210:198–210; 1966.

Dawson, W.R. Reptiles as Research Models in Comparative Physiology. *J. Am. J. Vet. Med.*, 159:1653–1661; 1971.
Derikson, W.K. Lipid Storage and Utilization in Reptiles. *Amer. Zool.*, 16:711–723; 1976.
Dessauer, H.C. Blood Chemistry of Reptiles: Physiology and Evolutionary Aspects. In: *Biology of the Reptilia, 3 C, Morphology* eds. C. Gans, and T.S. Parsons. 25–26. New York: Academic Press, 1970.
DiMaggio, A., III. Effects of Glucagon and Insulin on Carbohydrate Metabolism in a Lizard. *Fed. Proc. Fed. Am. Soc. Biol.*, 20:175; 1961.
Dunson, W.A. Salt Gland Excretion in the Pelagic Sea Snake, *Pelamis. Am. J. Physiol.*, 215:1512–1517; 1968.
———. Electrolyte Excretion by the Salt Glands of the Galapagos Marine Iguana. *Am. J. Physiol.*, 216:995–1002; 1969.
———. Reptilian Salt Glands. In *The Exocrine Glands* eds. S.Y. Botelho, et al. 89–103. Philadelphia: Univ. Pennsylvania Press, 1969.
———. Some Aspects of Electrolyte and Water Balance in Three Estuarine Reptiles, the Diamondback Terrapin, American and "Salt Water" Crocodiles. *Comp. Biochem. Physiol.*, 32:161–174; 1970.
——— et al. Sea Snakes: An Unusual Salt Gland Under the Tongue. *Science*, 173:437–441; 1971.
——— and Taub, A.M. Extra-Renal Salt Excretion in Sea Snakes (*Laticauda*). *Am. J. Physiol.*, 213:975–982; 1967.
Farnsworth, R.J., Brannian, R.E., Fletcher, K.C., and Klassen, S. A Vitamin E-Selenium Responsive Condition in a Green Iguana. *J. Zoo Anim. Med.*, 17:42–45; 1986.
——— and Weymouth, R.D. Active Uptake of Sodium by Soft-Shell Turtles (*Trionyx spinifer*). *Science*, 149: 67–69; 1965.
Fitch, H.S. Reproductive Cycles in Lizards and Snakes. *Misc. Pub. Univ. Kansas Nat. Hist.*, 52:1–247; 1970.
———. Reproductive Cycles in Tropical Reptiles. *Occas. Pap. Mus. Nat. Hist. Univ. Kansas*, 96:1–53; 1982.
Fowler, M.E. Metabolic Bone Disease. In: *Zoo and Wild Animal Medicine*, ed. M.E. Fowler, 55–76. Philadelphia: W.B. Saunders, Co., 1986.
———. Comparison of Respiratory Infection and Hypovitaminosis A in Desert Tortoises. In: *Pathology of Zoo Animals*. eds. R.J. Montali, and G. Migaki, G. 93–97. Washington, DC: Smithsonian Institution Press. 1980.
Fox, A.M. and Musacchia, X.J. Notes on the pH of the Digestive Tract of *Chrysemys picta. Copeia*, 1959:337–339; 1959.
Freeland, W.J. and Janzen, D.H. Strategies in Herbivory of Mammals: the Role of Plant Secondary Compounds. *Amer. Natur.*, 108:269–289; 1974.
Frye, F.L. The Role of Nutrition in the Successful Management of Captive Reptiles. *Proc. Calif. Vet. Med. Assoc. 86th Meeting and Scientific Seminar*, 5–20; 1974.
———. *Biomedical and Surgical Aspects of Captive Reptile Husbandry*, 1st ed. 23–61, Edwardsville, KS: Veterinary Medical Publishing Co., Inc. 1981.
———. Feeding and Nutritional Diseases. In *Zoo and Wild Animal Medicine*, 2nd ed. ed. M.E. Fowler, 139–151. Philadelphia: W.B. Saunders. 1986.

———. Care and Feeding of Some Invertebrates Kept as Pets or study animals. *Proc. C.V.M.A. 99th Annual Meeting and Scientific Seminar.* San Francisco, 271–304, 1987.

———. Vitamin A Sources, Hypovitaminosis A, and Iatrogenic Hypervitaminosis A in Captive Chelonians. In: *Current Veterinary Therapy,* Vol. X. ed. R.W. Kirk, 791–796. Philadelphia: W.B. Saunders Co. 1989.

———. *Biomedical and Surgical Aspects of Captive Reptile Husbandry,* 2nd Ed. ed. F.L. Frye. Melbourne, FL: R. E. Krieger Publ. Co., Inc. 1991.

———. *Captive Invertebrates: A Guide to Their Biology and Husbandry.* Melbourne, FL: R.E. Krieger Publ. Co., Inc. In Press. 1991.

——— and Calvert, C. Preliminary Information on the Nutritional Content of Mulberry Silk Moth (*Bombyx mori*) Larvae. *J. Zoo Wildl. Med.,* 20(1):73–75; 1989a.

———. The Feeding of *Bombyx mori,* as Prey Insects for Captive Lizards: A Quantum Improvement Over *Gryllus* spp., *Tenebrio molitor,* or *Galleria* spp. *Abstr. First World Congr. Herpetol.* Section S-4. Canterbury, England; Rutherford College, University of Kent. 14 September, 1989.

———. The Nutritional Content of Calcium-Supplemented Mulberry Silk Moth (*Bombyx mori*) Larvae. *HerpetoPathologica.* in press. 1991.

——— and Carney, J.D. Parathyroid Adenoma in a Tortoise. *Vet. Med./Sm. An. Clin.,* 70:582–584; 1975.

——— and Dutra, F.R. Hypothyroidism in Turtles and Tortoises. *Vet. Med./Sm. An. Clin.,* 69:990–993; 1974.

———. Articular Pseudogout in a Turtle, (*Chrysemys p. elegans*). *Vet. Med./Sm. An. Clin.,* 71:655–659; 1976.

——— et al. Spontaneous Diabetes in a Turtle. *Vet. Med./Sm. An. Clin.,* 71:935–939; 1976.

——— and Schelling, S.H. Steatitis in a Caiman. *Vet. Med/Sm. An. Clin.,* 68:143–145; 1973.

Gans, C. et al. Courtship, Mating and Male Combat in Tuatara, *Sphenodon punctatus.* *J. Herp.,* 18(2):194–197; 1984.

Gardner, A.S. The Calcium Cycle of Female Day-Geckos (*Phelsuma*). *Herpetology J.,* 1:37–39; 1985.

Goellner, R.R. Tuataras (*Sphenodon punctatus*) at St. Louis Zoo. *Acta Path. et Zoo Antverpiensia,* 78:319–324; 1984.

Goode, M. *Echis colorata* (Palestine Saw-Scaled Viper): Water Economy. *Herp. Rev.,* 14(4):120; 1983.

Groves, F. The Swallowing of Shed Skin by the Snake, *Lampropeltis getulus getulus* Linnaeus. *Herpetologica,* 20:128; 1964.

——— and Altimari, W. Keratophagy in the Slender Vine Snake, *Uromacer oxyrhynchus.* *Herp. Rev.,* 8(4):124; 1977.

——— and Groves, J.D. Keratophagy in Snakes. *Herpetologica,* 28:45–46; 1972.

Grzimek, B. *Grzimek's Animal Life Encyclopedia,* Vol. 6, Reptiles. New York; Van Nostrand Reinhold Co., 1972.

Haggag, G. et al. Hibernation in Reptiles. I. Changes in Blood Electrolytes. *Comp. Biochem. Physiol.,* 16:457–465; 1965.

———. Hibernation in Reptiles. II. Changes in Blood Glucose, Hemoglobin, Red Blood Cell Count, Protein and Non-Protein Nitrogen. *Comp. Biochem. Physiol.*, 17:335–339; 1966.

Hirth, H.F. Some Aspects of the Natural History of *Iguana iguana* on a Tropical Strand. *Ecology*, 44:613–615; 1963.

Holmes, W.N and McBean, R.I. Some Aspects of Electrolyte Excretion in the Green Turtle, *Chelonia mydas mydas. J. Exp. Biol.*, 41:81–90; 1964.

Honneger, R.E. Beitrage zur Biologie von *Hydrodinastes gigas* (*Cyclagris gigas*) in Terrarium. *Acta Zoo. et Path. Antverpiensia*, 78:237–244; 1984.

Howard, C.H. Maintenance and Breeding of the Mangrove Snake, (*Boiga dendrophila*) at Twycross Zoo. *Acta Zoo. et Path. Antverpiensia*, 78:272–275; 1984.

Huff, T.A. The Husbandry and Propagation of the Madagascar Ground Boa, (*Acanthrophis dumerili*) in Captivity. *Acta Zoo. et Path. Antverpiensia*, 78:255–269; 1984.

Ineich, I. and Gardner, A.S. Qualitative Analysis of the Development of Endolymphatic Sacs by a Gecko (*Lepidodactylus lugubris*) in French Polynesia. *J. Herp.*, 23(4):414–418; 1989.

Ippen, R. Considerations on the Comparative Pathology of Bone Diseases in Reptiles. *Zentralbl. Alleg. Path.*, 108: 424–434; 1965.

Iverson, J.B. Adaptations to Herbivory in Iguanine Lizards. In: *Iguanas of the World: Their Behavior, Ecology and Conservation*. eds. G.M. Burghardt and A.S. Rand. 60–76; Park Ridge, NJ: Noyes Publishing. 1982.

Jackson, C.G., Jr. et al. An Accelerated Growth and Early Maturity in *Gopherus Agassizi* (Reptilia, Testudines). *Herpetologica*, 32(2):139–145; 1976.

Jackson, D.C. Buoyancy Control in the Freshwater Turtle, *Pseudemys scripta elegans. Science*, 166:1649–1651; 1969.

Jackson, O.F. and Cooper, J.E. Nutritional Diseases. In: *Diseases of the Reptilia*, Vol. 2. eds. J.E. Cooper and O.F. Jackson, 409–428; London: Academic Press, 1981.

Jarrett, A. Comparison and Experimental Keratinisation. In: *The Physiology and Pathophysiology of the Skin*. Vol. 1. ed. A. Jarrett, 123–143, London: Academic Press, 1973.

———. The Action of Vitamin A on Adult Epidermis and Dermis. In: *The Physiology and Pathophysiology of the Skin*. Vol. 6, ed. A. Jarrett, 2059–2091, London: Academic Press, 1980.

Jenkins, N.K. and Simkiss, K. The Calcium and Phosphate Metabolism of Reproducing Reptiles with Particular Reference to the Adder (*Vipera berus*). *Comp. Biochem. Physiol.*, 26:865–876; 1968.

Jensen, H.B. and With, T.K. Vitamin A and Carotenoids in the Liver of Mammals, Birds, Reptiles and Man, with Particular Regard to the Intensity of the Ultraviolet Absorption and the Carr-Proce Reaction of Vitamin A. *Biochem. J.*, 33:1771–1786; 1939.

Jes, H. *Lizards in the Terrarium*. ed. F.L. Frye. Hauppauge, N: Barron's Educational Series, 1987.

Junquiera, L.C.U. et al. Reabsorptive Function of the Ophidian Cloaca and Large Intestine. *Physiol. Zool.*, 39:151–159; 1966.

Kästle, W. Kalk als Zusatznahrung für Eschsen. *Aquar. Terrar. Zeitschr.*, 15:62; 1962.
Keown, G.L. A Case of Keratophagy in *Lampropeltis getulus californiae* (Blainville). *J. Herp.*, 7:315–316; 1973
Khalil, F. Excretion in Reptiles. II. Nitrogen Constituents of the Urinary Concretions of the Oviparous Snake *Zamenis diadema*, Schlegel. *J. Biol. Chem.*, 172:101–103; 1948.
———. Excretion in Reptiles. III. Nitrogen Constituents of the Urinary Concretions of the Viviparous Snake, *Eryx thebaicus*, Reuss. *J. Biol. Chem.*, 172:105–106; 1948.
Kien, T. Captive Breeding of Cobras in Vietnam (Communicated to and Written by C. Gans). *Acta Zoo. et Path. Antverpiensia*, 78:215–217; 1984.
Kluge, A.G. Cladistic Relationships in the Gekkonoidea (Squamata, Sauria). *Misc. Pub. Mus. Zool. Univ. Michigan*, 173:1–54; 1987.
Kramer, D.C. Geophagy in *Terrapene ornata ornata* Agassiz. *J. Herp.*, 7(2):138–139; 1973.
Langerwerf, B. Techniques for Large-Scale Breeding of Lizards from Temperate Climates in Green-House Enclosures (Breeding many Species of Lizards in Captivity, Aiming the Maintenance of Populations of Each Species Outside their Natural Habitat). *Acta Zoo. et Path. Antverpiensia*, 78: 163–176; 1984.
Leloup, P. Different Aspects of the Breeding of Venomous Snakes in Large Scale. *Acta Zoo. et Path. Antverpiensia*, 78:177–198; 1984.
Loftin, H. and Tyson, E. Iguanas as Carrion Eaters. *Copeia*, 1965(4):515.
Losos, J. and Greene, H.W. Ecological and Evolutionary Implications of Diet in Monitor Lizards. *Biol. J. Linnean Soc.*, 35: 379-407; 1988.
Luttenberger, F. Die Zucht van *Eunectes murinus*. *Acta Zoo. et Path. Antverpiensia*, 78:245–254; 1984.
Malaret, L. and Fitch, H.S. Effects of Artificial Overfeeding and Underfeeding on Reproduction in Four Squamates. *Acta Zoo. et Path. Antverpiensia*, 78:77–84; 1984.
Matz, G. La Reproduction des Reptiles et les Facteurs de son Induction. *Acta Zoo. et Path. Antverpiensia*, 78:33–68; 1984.
McBee, R.H. and McBee, V.H. The Hindgut Fermentation in the Green Iguana, *Iguana iguana*. In: *Iguanas of the World: Their Behavior, Ecology and Conservation*. eds. G.M. Burghardt and A.S. Rand. 77–83, Park Ridge, NJ: Noyes Publishing, 1982.
McKeown, S. Management and Propagation of the Lizard Genus *Phelsuma*. *Acta Zoo. et Path. Antverpiensia*, 78:149–161; 1984.
Medica, P.A., Bury, R.B., and Luckenbach, R.A. Drinking and Construction of Water Catchments by the Desert Tortoise, *Gopherus agassizi*, in the Desert. *Herpetologica*, 36(4):301–304; 1980.
Mehrtens, J.M. *Living Snakes of the World in Color* New York; Sterling Publ. Co., 1987.
Mettler, F., Palmer, D., Rübel, A., and Isenbügel, E. Gehäuft Auftretende Fälle von Parakeratrosen mit Epithelablösung der Landschildkröten. *Verhand-*

lungsber. XXIV Int. Symp. Erkrankungen der Zootiere. Veszprem. 245–248; 1982.

Meylan, A. Spongivory in Hawksbill Turtles: A diet of Glass. Science. 239:393–395; 1988.

Miller, D.E. Rainwater Drinking by the Mangrove Water Snake, Nerodia fasciata compressicauda. Herp. Rev., 16(3): 71; 1985.

Miller, T.J. Corallus enhydris enhydris: Food (Ovophagy). Herp. Rev., 14(2):46–47; 1983.

Montgomery, G.G. ed. The Ecology of Arboreal Folivores. Washington, DC: Smithsonian Institution Press, 1978.

Moody, S. Endolymphatic Sacs in Lizards: Phylogenetic and Functional Considerations. (Abstr). Soc. Study Amphib. Reptiles/Herpetologists' League Annual Meeting, Salt Lake City, Utah, p. 80, 1983.

Murphy, J.B. and Mitchell, L.A. Miscellaneous Notes on the Reproductive Biology of Reptiles. 6. Thirteen Varieties of the Genus Bothrops (Reptilia; Serpentes; Crotalidae). Acta Zoo. et Path. Antverpiensia, 78: 199–214; 1984.

Mushinsky, H.R. Observations of the Feeding Habits of the Short-tailed Snake, Stilosoma extenuatum in Captivity. Herp. Rev., 15(3): 57–58; 1984.

Naulleau, G. and Detrait, J. Incidence de l'Elevage en Captivite sur la Fonction Venimeuse chez Vipera aspis et Vipera ammodytes. Acta Zoo. et Path. Antverpiensia, 78: 219–236; 1984.

Norris, K.S. and Dawson, W.R. Observations on the Water Economy and Electrolyte Excretion of Chuckwallas (Lacertilia, Sauromalus). Copeia, 1964(4):638–646; 1964.

Oguro, C. and Sasayama, U. Morphology and Function of the Parathyroid Gland of the Caiman, Caiman crocodilus. Gen. Comp. Endocrinol., 19(2):161–169; 1976.

Packard, M.J., Packard, G.C. and Boardman, T.J. Structure of Eggshells and Water Relations of Reptilian Eggs. Herpetologica, 38:136–155; 1982.

——— and Gutzke, W.H.N. Calcium Metabolism of the Oviparous Snake Coluber constrictor. J. Exp. Biol, 110:99–112; 1984.

——— and Miller, J.D., et al. Calcium Mobilization, Water Balance, and Growth in Embryos of the Agamid Lizard, Amphibolurus barbatus. J. Exp. Biol., 235:349–357; 1985.

Pallaske, G. Hypervitaminosis D in a Lizard. Berl. Tierzarztl. Wchnschr., 74:132; 1961.

Palmer, D.G., Rübel, A., Mettler, F., and Volker, L. Experimentell erzeugte Hautveränderungen bei Landschildkröten durch hohe parenterale Gaben von Vitamin A. Zbl. Vet. Med., A. 31:625–633; 1984.

Paul-Murphy, J, Mader, D.R., Kock, N., and Frye, F.L. Necrosis of Esophageal and Gastric Mucosa in Snakes Given Oral Dioctyl Sodium Sulfosuccinate. Proc. A.A.Z.V., 474–477; 1987.

Peaker, M. Active Acquisition of Stomach Stones in the American Alligator, Alligator mississippiensis Daudin. Brit. J. Herpetol., 4:103–104; 1969.

——— Some Aspects of the Thermal Requirements of Reptiles in Captivity. Internat. Zoo Yearbook, 9:3–8; 1969.

Pfeiffer, C. Foods for Tortoises, I–X. *Turtle Hobbyist*, 1,2; 5163 E. Bedford Drive, San Diego, CA 92116. Undated.

Porter, K.R. *Herpetology*. Philadelphia: W.B. Saunders Co., 1972.

Pritchard, P.C.H. A Reinterpretation of *Testudo gigantea* Schweigger, 1812. *J. Herpetology*, 20 (4):522–534; 1986.

Rand, A.S. Reptilian Arboreal Folivores. In: *The Ecology of Arboreal Folivores*. ed. G.G. Montgomery, 115–122. Washington, DC: Smithsonian Institution Press, 1978.

——, Dugan, B.A., Monteza, H., and Vianda, D. The Diet of a Generalized Folivore: *Iguana iguana* in Panama. *J. Herp.*, 24(2):211–214. 1990.

Redford, K.H. and Dorea, J.G. The Nutritional Value of Invertebrates with Emphasis on Ants and Termites as Food for Mammals. *J. Zool. London*, 203:385–395, 1984.

Reichenbach-Klinke, H.H. and Elkan, E. *The Principal Diseases of Lower Vertebrates*. New York: Academic Press, 1965.

Rhodin, A.G.J. Pathological Lithophagy in *Testudo horsfieldi*, *J. Herp.*, 8:385–386; 1974.

Rhoten, W.B. Glucagon Levels in Pancreatic Extracts and Plasma of the Lizard. *Amer. J. Anat.*, 147(1):131–138; 1976.

Roots, C. The Chemical Composition of the Mealworm. *WRC Wildlife J.* 1(4), 1978

Rost, D.R. and Young, M.C. Diagnosing White-Muscle Disease. *VM/SAC* 80:1286–1287; 1984.

Ruth, E.S. A Study of the Calcium Glands in the Common Phillipine House Lizards. *Phil. J. Sci. (B) Trop. Med.*, 13:311–319;1918.

Saint-Girons, H. Regime et Rations des Serpents. *Bull. de la Societe Zoologique de France*, 108(3):431–437; 1983.

Savidge, J.A. Food Habits of *Boiga irregularis*, an Introduced Predator on Guam. *J. Herpetol.*, 22(3):275–282; 1988.

Sazima, I. Feeding Behavior of the Snail-eating Snake, *Dipsas indica*. *J. Herp.*, 23(4):464–468; 1989.

Schmidt-Nielsen, K. *Desert Animals*. New York: Oxford University Press, 225–251, 1964.

—— and Frange, R. Salt Glands in Marine Reptiles. *Nature*, (London, England) 182:783–785; 1958.

—— et al. Nasal Salt Excretion and the Possible Function of the Cloaca in Water Conservation. *Science*, 142:1300–1301; 1963.

Schuchman, S.M. and Taylor, D.O.N. Arteriosclerosis in an Iguana (*Iguana iguana*). *J. Am. Vet. Med. Assoc.*, 157:614–616; 1970.

Seidel, M.E. and Smith, H.M. Chrysemys, Pseudemys, Trachemys (Testudines:Emydidae): Did Agassiz Have it Right? *Herpetologica*, 42(2):242–248; 1987.

Shine, R. Food Habits and Reproductive Biology of Australian Elapid Snakes of the Genus *Denisonia*. *J. Herp.* 17(2):171–175; 1983.

Simkiss, K. *Calcium in Reproduction Physiology*. London: Chapman and Hall, 1967.

Skorepa, A.C. The Deliberate Consumption of Stones by the Ornate Box Turtle, *Terrapene ornata* Agassiz. *J. Ohio Herp. Soc.*, 5:108; 1966.

Smith, H.A. et al. *Veterinary Pathology*, 4th ed. 1057–1071. Philadelphia: Lea & Febiger, 1972.

Sokol, O.M. Herbivory in Lizards. *Evolution*, 21:192–194; 1967.

———. Lithophagy and Geophagy in Reptiles. *J. Herp.*, 5:69–71; 1971.

Sprackland, R. Feeding and Nutrition of Monitor Lizards in Captivity and in the Wild. *Bull. Kansas Herpetol. Soc.*, 47:15–18; 1982.

———. A Preliminary Study of the Food Discrimination in Monitor Lizards (Reptilia: Lacertilia: Varanidae). *Bull. Chicago Herpetol. Soc.*, 25(10):181–183, 1990.

Studer, A. and Frey, J.R. Über Hautveränderungen der Rattenach Groffen Oralen Dosen von Vitamin A. *Schweiz. Med. Wschr.*, 79:382–384; 1949.

Sylber, C.K. Feeding Habits of the Lizards *Sauromalus varius* and *S. hispidus* in the Gulf of California. *J. Herp.*, 22(4):413–424; 1988.

Szarski, H. Some Remarks on Herbivorous Lizards. *Evolution*, 16:529; 1962.

Taub, A.M. and Dunson, W.A. The Salt Gland in a Sea Snake, (*Laticauda*). *Nature*, 215:995–996; 1967.

Templeton, J.R. Nasal Salt Excretion in Terrestrial Lizards. *Comp. Biochem. Physiol.*, 11:223–229; 1964.

Throckmorton, G.S. Oral Food Processing in Two Herbivorous Lizards, *Iguana iguana* (Iguanidae) and *Uromastix aegyptius* (Agamidae). *J. Morph.*, 148:363–390; 1976.

Tonge, S. and Bloxam, Q. Captive Reproduction of Rhinoceros Iguana (*Cyclura c. cornuta*) in Indoor Accommodation. *Acta Zoo et Path. Antverpiensia*, 78:115–128; 1984.

Troyer, K. Transfer of Fermentative Microbes Between Generations in a Herbivorous Lizard. *Science*, 216:540–542; 1982.

———. Structure and Function of the Digestive Tract of Herbivorous Lizard, *Iguana iguana*. *Physiol. Zool.*, 57(1):1–8;1984.

Tucker, V.A. The Role of the Cardiovascular System In Oxygen Transport in Lizards. In: *Lizard Ecology, a Symposium*. ed. W.W. Milstead, 258–269. Columbia, MO: University of Missouri Press, 1967.

Van Devender, R.W. Growth and Ecology of Spiny-Tailed and Green Iguanas in Costa Rica, with Comments on the Evolution of Herbivory and Large Body Size. In: *Iguanas of the World: Their Behavior, Ecology and Conservation*. eds. G.M. Burghardt and A.S. Rand, 162–183. Park Ridge, NJ: Noyes Publishing. 1982.

Van Vleet, J.F. Comparative Pathology of Selenium and Vitamin E Deficiency and Excess. *Comparative Path. Bull.*, 17(4):1,4; 1985.

Visser, G. Husbandry and Reproduction of the Sail-Tailed Lizard (*Hydrosaurus amboinensis*) (Reptilia; Sauria; Agamidae) at Rotterdam Zoo. *Acta Zoo. et Path. Antverpiensia*, 78:129–148; 1984.

Vogel, P. Hettrich, W., and Ricono, K. Weight Growth of Juvenile Lizards, *Anolis lineatopus*, Maintained on Different Diets. *J. Herp.*, 20(1):50–58; 1986.

Vosburgh, K.M., Brady, P.S. and Ullrey, D.E. Ascorbic Acid Requirements of Garter Snakes: Plains (*Thamnophis radix*) and Eastern (*T. sirtalis sirtalis*). *J. Zoo Anim. Med.*, 13:38–42; 1982.

Wallach, J.D. Hypervitaminosis D in Green Iguanas. *J. Amer. Vet. Med. Assoc.*, 149:912–914; 1966.

———. Environmental and Nutritional Diseases of Captive Reptiles. *J. Am. Vet. Med. Assoc.*, 159:1632–1643; 1971.

———. Steatitis in Captive Crocodilians. *J. Amer. Vet. Med. Assoc.*, 153:845–847; 1968.

——— and Hoessle, C. Fibrous Osteodystrophy in Green Iguanas. *J. Am. Vet. Med. Assoc.*, 153:863–865; 1968.

Werner, Y.L. Observations on the Eggs of Eublepharid Lizards, with Comments on the Evolution of the Gekkonoidea. *Zool. Med.*, 47:211–224, 1972.

Whitaker, R. Captive Breeding of Crocodilia in India. *Acta Zoo. et Path. Antverpiensia*, 78: 309–318, 1984.

Whitehair, C.K. White Muscle Disease. In: *Bovine Medicine & Surgery*, 1st Ed. Santa Barbara: American Veterinary Publications. 1970.

Whiteside, B. The Development of the Saccus Endolymphaticus in *Rana temporaria* Linne. *Amer. J. Anat.*, 30:231–266; 1922.

Zain, B.K. and Zain-ul-Abdelin, M. Characterization of the Abdominal Fat Pads of a Lizard. *Comp. Biochem. Physiol.*, 23:173–177; 1967.

Zaworski, J.P. A Note of Calcium Supplementation in Geckos. *Bull. Chi. Herp. Soc.*, 22(8):131, 1987.

———. Ophiophagy in *Pituophis melanoleucus deserticola*. *Bull. Chicago Herp. Soc.*, 25(6):100, 1990.

Zwart, P. and Van de Watering, C.C. Disturbances of Bone Formation in the Common Iguana (*Iguana iguana* L.): Pathology and Etiology. *Acta Zoo et Path. Anat.*, 48:333–356; 1969.

Appendix D

Species List Cross-Referenced by Common Name

Please note: where a particular species may be known by more than one common name, the other names are listed.

adder; crossed viper; kreuzotter	*Vipera berus*
Aesculapian snake	*Elaphe l. longissima*
African beaked snake	*Rhamphiophis multimaculatus*
African helmeted turtle	*Pelomedusa subrufa*
African house snake	*Lamprophis fulginosus*
African mole snake	*Pseudaspis c cana*
African rock python	*Python sebae*
African side-neck turtles	*Pelusios sp.*
agama lizards	*Agama sp.*
Aldabra tortoise	*Geochelone (Testudo) gigantea*
alligator, American	*Alligator mississippiensis;*
alligator, Chinese	*Alligator sinensis*
alligator lizards	*Gerrhonotus sp.; Elgaria sp.*
alligator snapping turtle	*Macrochelys temminckii*
amethystine python	*Liasis amethystinus*
Amur viper	*Agkistrodon intermedius*
anaconda	*Eunectes murinus; E. notaeus*
Angolan python	*Python anchietae*
anole lizards	*Anolis sp.*
Argentine green speckled snake	*Leimadophis poecilogyrus*
Argentine side neck turtle	*Phrynops hilarii*
armadillo lizards	*Cordylus sp.*
Aruba rattlesnake	*Crotalus unicolor*
Asian big tooth snake	*Didodon rufozonatum*

Asian box turtles	*Cuora* sp.
Asian rat snake	*Gonyosoma oxycephala*
asp	*Vipera aspis*
Australian tree snake	*Dendrelaphis punctulatus*
Baja bullsnake	*Pituophis m. vertebralis*
Baja California rattlesnake	*Crotalus enyo*
ball python	*Python regius*
bamboo viper; Chinese tree viper	*Trimeresurus stejnegeri*
banded gecko	*Cyrtodactylus pulchellus*
banded krait	*Bungarus fasciatus*
bandy-bandy snake	*Vermicella annulata*
barba amarilla; fer-de-lance snake	*Bothrops andianus asper*
Barbour's pit viper	*Porthidium (Bothrops) barbouri*
basilisk lizards	*Basiliscus* sp.
beaded lizard, Mexican	*Heloderma horridum*
beaked snake	*Rhamphiohis multimaculatus*
bearded lizard	*Amphibolurus barbatus*
beauty snake	*Elaphe taeniua friesi*
Berlander's (Texas) tortoise	*Gopherus (Xerobates) berlandieri*
big-headed turtle	*Platemys megacephalum*
bird snake	*Thelotornis kirtlandii*
Bismark ringed python	*Liasis boa*
black-headed python	*Aspidites melanocephalus*
black-headed snakes	*Tantilla* sp.
black pine snake	*Pituophis m. lodingi*
black pond turtle	*Seibenrockiella crassicollis*
black rat snake	*Elaphe obsoleta*
black rattlesnake	*Crotalus viridis cerberus*
black striped snake	*Coniophanes imperalis*
black swamp snake	*Seminatrix p. pygaea*
black-tailed horned pit viper	*Porthidium (Bothrops) melanurus*
black-tailed rattlesnake	*Crotalus molossus*
black tree monitor lizard	*Varanus beccari*
black tree snake	*Thasops jacksoni*
Blanding's turtle	*Emydoidea blandingi*
blind snake	*Rhamphotyphlops* sp.
blood python	*Python curtus*
blue krait	*Bungarus caeruleus sindanus*
blue-striped garter snake	*Thamnophis s. similis*
blue-tongued skinks	*Tiliqua* sp.
blunt-headed tree snake	*Imantodes cenchoa*
boas, common	*Boa constrictor* ssp.
boas, dwarf, Caribbean Island	*Exiliboa, Tropidophis, Ungaliophis*
boa, rainbow	*Epicrates,*
boas, rosy	*Lichanura* sp.
boa, rough-skinned dwarf	*Trachyboa,*

boa, rubber	*Charina bottae*
boa, sand	*Eryx sp.*
boa, tree	*Tropidophis*
Boelen's python	*Liasis boeleni*
bog turtle	*Clemmys muhlenbergi*
Bolson tortoise	*Gopherus (Xerobates) flavomarginata*
boomslang	*Dispholidus typus*
box turtle	*Terrapene sp.*
Brahminy River turtle	*Hardella thurgi*
Brazilian smooth snake	*Cyclagras gigas*
brown tree snake	*Boiga irregularis*
brown water python	*Liasis fuscus*
bull snake	*Pituophis melanoleucus spp.*
Burmese mountain tortoise	*Geochelone emys*
Burmese python	*Python molurus bivittatus*
bushmaster	*Lachesis muta*
Butler's garter snake	*Thamnophis butleri*
caimans	*Caiman, sp.; Melanosuchus niger*
caiman lizard	*Dracaena guianensis*
Calabar burrowing python	*Calabaria reinhardtii*
California kingsnake	*Lampropeltis getulus californiae*
Canary Island giant lacerta	*Gallotia stelini*
canebrake rattlesnake	*Crotalus atriocaudatus*
carpet python	*Morelia spilotes variegata*
cascavel, cascabel rattlesnake	*Crotalus durissus*
cat-eyed snake	*Leptodeira septentrionalis;*
cat-eyed snake; herald snake	*Crotaphopeltis botamboeia*
Chaco tortoise	*Chelonoides chilensis*
chameleons, old world	*Chamaeleo sp.; Brooksia sp.*
checkered gartersnake	*Thamnophis marcianus*
chicken turtle	*Deirochelys reticularia*
Children's python	*Morelia childreni*
Chinese king rat snake	*Elaphe carinata*
Chinese lined rat snake	*Elaphe rufodorsata*
Chinese rat snake	*Elaphe taeniua friezi*
Chinese tree viper	*Trimeresurus stejnegeri*
Chinese twin spot rat snake	*Elaphe bimaculata*
chondropython	*Chondropython viridis*
chuckwalla lizard	*Sauromalus obesus; S. varius*
coachwhip snake	*Masticophis*
cobras	*Aspidelaps lubricus; Boulengerina annulata; Hemachatus hemachatus; Naja sp; Pseudohaje goldii; Walterinnesia sp.; Ohiophagus hannah*

collared lizards	*Crotaphytus* sp.
common garter snake	*Thamnophis sirtalis*
cooter turtle	*Chrysemys*
coral snakes	*Micrurus* sp.; *Micuroides* sp.; *Maticora birvirgata flaviceps*
copperhead rat snake	*Elaphe flavolineata*
copperhead snake	*Agkistrodon contortrix*
crawfish (or swamp) snakes	*Regina* sp.
cribo, yellow-tailed cribo	*Drymarchon c. corais*
crocodiles	*Crocodylus* sp.
crocodile monitor lizard	*Varanus salvadori*
crowned snakes	*Tantilla* sp.
old world smooth crowned snake	*Coronella girondica*
Cuban boa	*Epicrates angulifer*
curly-tailed lizard	*Leiocephalus carinatus*
dab lizard	*Uromastix acanthinurus*; *U. aegypticus*
death adder	*Acanthophis antarcticus*
DeKay's snake	*Storeria dekayi*
desert kingsnake	*Lampropeltis getulus splendida*
desert tortoise	*Gopherus (Xerobates) agassizi*
desert viper	*Cerastes cerastes*
diadem snake	*Spalerosophis diadema cliffordi*
diamondback terrapin	*Malaclemys terrapin*
dice snake	*Natix t. tessellata*
dusky rattlesnake	*Crotalus pusillus*
dwarf sand adders	*Bitis peringueyi*; *B. schneideri*
dwarf sand lizard	*Eremias grammica*
dwarf tegu	*Calliopistes maculatus*
earless lizards	*Holbrookia* sp.; *Cophosaurus* sp.
earth snake, rough	*Virginia striatula*
earth snake, smooth	*Virginia valeriae*
East African side neck turtles	*Pelusios* sp.
eastern diamondback rattlesnake	*Crotalus adamanteus*
eastern ribbon snake	*Thamnophis s. sauritus*
eastern kingsnake	*Lampropeltis g. getulus*
egg-eating snake	*Dasypeltis scabra*; *D. atra*
Egyptian tortoise	*Testudo kleinmanni*
emerald tree boa	*Corallus canina*
elephant trunk snake (or Kurung)	*Acrochordus* sp.
elongated tortoise	*Geochelone elongata*
eye lash viper	*Bothriechis (Bothrops) schlegeli*
false coral snakes	*Erythrolamprus bizona*; *Lampropeltis* sp.
false habu snake	*Macropisthodon rudis*
false map turtle	*Graptemys pseudogeographica*

false water cobra	*Cyclagras gigas*
fat-footed gecko	*Ptyodactylus hasselquistii*
fat-tailed gecko	*Hemitheconyx caudicinctus*
Fea's viper	*Azemiops feae*
fence lizards	*Sceloporus sp.*
fer-de-lance snake	*Bothrops andianus asper*
fishing snake; tentacled snake	*Erpeton tentaculum*
flat-headed snakes	*Tantilla sp.*
Florida kingsnake	*Lampropeltis getulus floridana*
Fly River turtle	*Carettochelys insculpta*
flying gecko	*Ptychozoon kuhli*
flying snakes	*Chrysolopea sp.*
forest racer	*Drymoluber dichrous*
four-eyed day gecko	*Phelsuma quadriocellata*
four-eyed turtle	*Sacalia bealei*
fox snake	*Elaphe vulpina*
fringe-toed lizards	*Uma sp.*
frog-eyed gecko	*Teratoscincus S. scincus*
Gaboon viper	*Bitis g. gabonica*
Galapagos tortoises	*Geochelone elephantopus ssp.*
garter snakes	*Thamnophis sp.*
gavial (gharial)	*Gavialis gangeticus; Tomistoma schlegeli*
geckos	*Chondrodactylus sp.; Coelonyx sp.; Eublepharis sp.Gekko sp.; Gehyra sp.; Gonatodes sp.; Hemidactylus sp.; Phelsuma sp. Phyllodactylus sp.; Pytodactylus sp; Shaerodactylus sp.; Tarentola sp., etc.*
Gila monster	*Heloderma suspectum*
glass "snake" lizard	*Ophiosaurus ventralis*
glossy snake	*Arizona elegans.*
Godman'a pit viper	*Porthidium (Bothrops) godmani*
gold thread turtle	*Ocadia sinensis*
Gold's tree cobra	*Pseudohaje goldii*
gopher snake (bull snake)	*Pituophis melanoleucus ssp.*
gopher tortoise	*Gopherus polyphemus*
gray-banded kingsnake	*Lampropeltis mexicana alterna*
gray rat snake	*Elaphe obsoleta spilotes*
Gray's monitor lizard	*Varanus olivaceus (V. grayi)*
Great Basin gopher snake	*Pituophis m. deserticola*
Great Plains rat snake	*Elaphe obsoleta emoryi*
Great Plains gopher snake	*Pituophis melanoleucus deserticola*
Greek tortoise	*Testudo graeca*
green headed tree snake	*Chironius scurrulus*

green night adder	*Causus resimus*
green ratsnake	*Elaphe triaspis intermedia*
green snake, rough	*Opheodrys vernalis*
green snake, smooth	*Liopeltis sp.*
green tree monitor lizard	*Varanus praesinus*
green tree python	*Chondropython viridis*
green turtle, marine	*Chelonia mydas*
Greer's kingsnake	*Lampropeltis mexicana greeri*
habu snakes	*Trimeresurus sp.*
hawksbill turtle, marine	*Eretmochelys imbricata*
helmeted lizard	*Corythophanes sp.*
herald snake; cat-eyed snake	*Crotaphopeltis botamboeia*
Hermann's tortoise	*Testudo hermanni*
himehabu	*Trimeresurus okinavensis*
hingeback tortoises	*Kinexys sp.*
hog-nosed snakes	*Heterodon platyrhinos ssp.; H. nasicus; H. simus; Lioheterodon madagascariensis*
hog-nosed snake, South American	*Lystrophis sp.*
hog-nosed viper	*Porthidium (Bothrops) nasutus*
hook-nosed snake	*Glyalopion sp., Ficimia streckeri*
horned adder	*Bitis caudalis*
horned desert viper	*Cerastes c. cerastes*
hornless desert viper	*Cerastes c. gasperetti; Cerastes vipera*
house gecko	*Hemidactylus garnoti*
iguana, common	*Iguana iguana*
iguana, desert	*Dipsosaurus dorsalis*
iguana, Fiji Island	*Brachylophus sp.*
iguana, ground	*Cyclura sp.; Conolophus pallidus*
iguana, marine	*Amblyrhynchus*
iguanas, spiny-tailed	*Ctenosaurus sp.*
impressed tortoise	*Geochelone impressa*
Indian rat snake	*Elaphe helena*
Indian rock python	*Python molurus*
Indian tree viper	*Trimeresurus gramineus*
indigo snake	*Drymarchon corais ssp.*
Indonesian green water snake	*Enhydris plumbea*
Japanese rat snake	*Elaphe climacophora*
jumping viper	*Porthidium (Bothrops) mummifer mexicanum*
jungle runner lizard	*Ameiva ameiva*
Kanburian pit viper	*Trimeresurus kanburiensis*
keeled rat snake	*Zaocys dhumnades*
kingsnakes	*Lampropeltis sp.*
king cobra	*Ophiophagus hannah*

Kirtland's bird snake	*Thelotornis kirtlandii*
Kirtland's snake	*Clonophis kirtlandi*
kraits	*Bungarus* sp.
Kaznakov's viper	*Vipera kaznakovi*
lacerta lizards, misc.	*Lacerta* sp.
lance-headed rattlesnake	*Crotalus polystictus*
leaf-nosed snake	*Phyllorhynchus decurtatus*
leaf-nosed viper	*Eristocophis macmahoni*
leaf-tailed gecko	*Phelsuma serratocauda*
leaf turtles	*Cyclemys* sp.
leatherback turtle, marine	*Dermochelys coriacea*
legless lizard	*Anniella pulcra*
leopard gecko	*Eublepharis macularius*
leopard lizards	*Crotaphytus collaris* ssp.
leopard tortoise	*Geochelone pardalis* ssp.
lined snake	*Tropidoclonium lineatum*
loggerhead turtle, marine	*Carretta carretta*
long-headed rattlesnake	*Crotalus polystictus*
long-nosed viper	*Vipera ammodytes*
long-tailed brush lizard	*Urosaurus graciosus*
long-tailed rattlesnake	*Crotalus stejnegeri*
Louisiana pine snake	*Pituophis m. ruthveni*
lyre snake	*Trimorphodon biscutatus*
mahogany rat snake	*Pseutes poecilonotus polylepis*
Malagasy hognose snake	*Lioheterodon madagascarensis*
Malagasy tree boa	*Sanzinia madagascariensis*
Malayan long-glanded coral snake	*Maticora bivirgata flaviceps*
Malayan pit viper	*Calloselasma rhodosotma*
mambas	*Dendroaspis* sp.
mamushi snake	*Agkistrodon blomhoffi*
mangrove monitor	*Varanus indicus*
mangrove pit viper	*Trimeresurus purpureomaculatus*
mangrove snake	*Boiga dendrophila*
many-banded krait	*Bungarus m. multicinctus*
many-horned adder	*Bitis cornuta*
map turtles	*Graptemys geographica* ssp.
marine (or sea) snakes	*Laticauda* sp.; *Pelamis* sp.
massasaugas	*Sistrusus catenatus* spp.; *S. ravus*
mata mata turtle	*Chelys fimbriata*
Mexican milksnake	*Lampropeltis triangulum annulata*
Mexican moccasin, cantil	*Agkistrodon bilineatus*
Mexican ratsnake	*Elaphe flavirufa*
Mexican green rattlesnake	*Crotalus basiliscus*
milksnake	*Lampropeltis triangulum*
Moelendorff's rat snake	*Elaphe moellendorffi*
Mojave rattlesnake	*Crotalus scutulatus*

mole snake, African	*Pseudaspis c. cana*
moloch lizard	*Moloch horridus*
monitor lizards	*Varanus* sp.
Montpellier snake	*Malopon m. monspessulanus*
mountain adder	*Bitis atropos*
mountain horned lizard	*Acanthosaura armata*
mud snake	*Farancia abacura*
mud turtles	*Kinosternon* sp.
Muhlenberg's turtle	*Clemmys muhlenbergi*
musk turtles	*Sternotherus* sp.
mussurana	*Clelia clelia*
neck-banded snake	*Scaphiodontophis annulatus bondurensis*
New Guinea rat snake	*Elaphe flavolineata*
New Guinea side neck turtle	*Elysea novaguineae*
New Guinea tree boa	*Candoia carinata*
night adders	*Causus* sp.
night lizards	*Xantusia* sp.
night snakes	*Eridiphas* sp.; *Hypsiglena* sp.
Nile monitor lizard	*Varanus niloticus*
northern Pacific rattlesnake	*Crotalus viridis oreganus*
northern pine snake	*Pituophis m. melanoleucus*
olive python	*Python olivaceus*
olive ridley turtle, marine	*Lepidochelys olivacea*
painted turtle	*Chrysemys picta* ssp.
Palestinian viper	*Vipera palaestinae*
palm vipers	*Bothriechis (Bothrops)* sp.
pancake tortoises	*Malacochersus tornei; M. procteri*
parrot snakes; flying snakes	*Leptophis ahaetulla; L. mexicana; Chrysolopea* sp.
patch-nosed snake	*Salvadora hexalepis*
peach throat monitor	*Varanus karlschmidti*
peninsula ribbon snake	*Thamnophis s. sackeni*
Peroni's sea snake	*Acalyptophis peronii*
Philippine rat snake	*Elaphe erythura*
pine snakes	*Pituophis melanoleucus* spp.
pipe snakes	*Cylindrophis* sp.
pit vipers, Asian	*Agkistrodon* sp.; *Calloselasma rhodostoma; Deinagkistrodon acutus; Trimeresurus* sp.
pit vipers; new world	*Agkistrodon* sp.; *Bothrops* sp.; *Crotalus* sp.; *Lachesis muta; Ophryacus undulatus; Porthidium* sp.; *Sistrurus* sp.
plated lizards	*Gerrhosaurus* sp.; *Zonosaurus* sp.
pond turtle	*Clemmys* sp.

Pope's tree viper	*Trimeresurus popeorum*
prairie kingsnake	*Lampropeltis calligaster*
prehensile-tail skink	*Corucia zebrata*
puff adder	*Bitis arietans*
puffing snakes	*Pseutes poecilonotus polylepis*; *P. sulphureus*
pygmy rattlesnake	*Sistrurus miliarius*
queen snake	*Regina septemvittata*
racer, blue	*Coluber constrictor* ssp.
racer, Ravergier's	*Haemorrhois ravergieri*
racer, West Indian	*Alsophis vudii picticeps*
racer, yellow-bellied	*Coluber constrictor mormon*
radiated snake	*Elaphe radiata*
radiated tortoise	*Testudo radiata*
rainbow boa	*Epicrates cenchria*
rainbow lizard	*Ameiva ameiva*
rainbow snake	*Farancia e. erythrogramma*
rat (or chicken) snakes	*Elaphe* sp., *Spilotes*, sp., *Zaocys dhumnades*
rattlesnakes	*Crotalus* sp.
red-bellied snake	*Storeria occipitomaculata*
red diamond rattlesnake	*Crotalus ruber*
red-eared slider turtle	*Trachemys scripta elegans*
red-footed tortoise	*Geochelone carbonaria*
red milksnake	*Lampropeltis triangulum syspila*
red rat snake	*Elaphe g. guttata*
red-necked keelback snake	*Rhabdophis subminiatus*
red-tailed pipe snake	*Cylindrophis rufus*
Reeve's turtle	*Chinemys reevesi*
reticulated python	*Python reticulatus*
rhinoceros viper	*Bitis nasicornis*; *Bitis gabonica rhinoceros*
ribbon snakes	*Thamnophis* sp.
ridge-nosed rattlesnakes	*Crotalus willardi*
ringed sawback turtle	*Graptemys oculifera*
river turtle, Central American	*Dermatemys mawei*
rock lizard, banded	*Petrosaurus mearnsi*
rock rattlesnakes	*Crotalus lepidus* ssp.
rosy boa	*Lichanura trivirgata*
rough necked monitor lizard	*Varanus rudicollis*
rough scaled boa	*Trachyboa boulengeri*
rough scaled sand boa	*Gongylophis conicus*
ruin lizard	*Podarcis sicula*
Russell's viper	*Vipera russelli*
Russian gargoyle lizard	*Phrynocephalus mystaceus*
Russian rat snake	*Elaphe schrenckii*

sage-brush	*Sceloporus* sp.
sail-finned; sail-tailed lizard	*Hydrosaurus* sp.
San Francisco garter snake	*Thamnophis s. tetrataenia*
sand adder	*Vipera ammodytes*
sand "fish" (skink)	*Scincus scincus*
sand lizard (gecko)	*Acanthodactylus pardalis*
sand skink	*Chalcides sexlineatus*
sand snake	*Psammophis* sp.
savannah monitor	*Varanus exanthematicus*
saw scaled viper	*Echis carinatus* ssp.
scarlet kingsnake	*Lampropeltis triangula elapsoides*
scarlet snake	*Cemophora c. coccinea*
sea snakes	*Acalyptophis peronii*; *Aipysurus duboisi*; *A. eydouxi*; *Astrotia stokesii*; *Emydocephalus annulatus*; *Hydrophis* sp.; *Lapemis hardwicki*; *Laticauda* sp.; *Pelamis platurus*
sharp-nosed viper	*Deinagkistrodon acutus*
sharp-tailed snake	*Stilosoma extenuatum*
sheltopusik lizard	*Ophiosaurus apodus*
shovel-nosed snake	*Chionactes* sp.
Siamese palm viper; Wirot's pit viper	*Trimeresurus wirotii*
side-blotched lizards	*Uta stansburiana*; *U. palmeri*
side-necked turtle	*Chelodina longicollis*, etc.
sidewinder rattlesnake	*Crotalus cerastes*
Sinai Desert cobra	*Walterinnesia aegyptia*
skinks, large, Australasian, (fruit eating)	*Chalcides* sp.; *Corucia* sp.; *Egernia* sp.; *Mabuya*; sp.; *Tiliqua* sp.; *Trachydosaurus* sp.
skinks, small, insectivorous	*Eumeces* sp.; *Scincella*; sp. *Scincus* sp.; *Lerista* sp.
slider turtles	*Trachemys*
slow "worm" lizard	*Anguis fragilis*
small-headed rattlesnake	*Crotalus intermedius gloydi*
smooth green snake	*Opheodrys* sp.
smooth scaled water snake	*Enhydris plumbea*
snake eating turtle	*Cuora flavomarginata*
snake neck turtles	*Chelodina* sp.
snail-eating snake	*Dipsas indica, D.* sp.; *Tropidodipsas sartori*
snapping turtle, alligator	*Macrochelys temminckii*
snapping turtle, common	*Chelydra serpentina*
soft-shelled turtles	*Trionyx* sp.
Solomon Island ground boa	*Candoia carinata*; *C. bibroni*

Sonoran gopher snake	*Pituophis melanoleucus affinis*
South American wood turtle	*Rhinoclemmys punctularia* spp.
southern pine snake	*Pituophis m. mugitus*
southern Pacific rattlesnake	*Crotalus viridis helleri*
speckled kingsnake	*Lampropeltis getulus holbrooki*
speckled rattlesnake	*Crotalus mitchelli pyrrhus*
speckled snake, Argentine	*Leimadophis poecilogyrus reticulatus*
spiny lizard	*Sceloporus magister*
spiny neck turtle	*Platemys spixii*
spiny-tailed iguanas	*Ctenosaura pectinata*; *Urocentron* sp.
spiny-tailed lizards	*Uromastix* sp.
spiny turtle	*Geomyda spinosa*
spitting cobra	*Naja n. sputatrix*
spotted turtle	*Clemmys guttata*
spurred tortoise, African	*Geochelone sulcata*
Sri Lanken pit viper	*Trimeresurus trigonocephalus*
star tortoise	*Testudo elegans*
stinkpot turtle	*Sternotherus odoratus*
Sumatran tree viper	*Trimeresurus sumatranus*
sunbeam snake	*Xenopeltis unicolor*
swifts, North American	*Sceloporus* sp.
swifts, Latin American	*Liolaemus* sp.
taipan snake	*Oxyuranus scutellatus*
Taiwan beauty snake	*Elaphe t. friesi*
Taiwan kukri snake	*Oligodon formosanus*
tegu lizards	*Tupinambis nigropunctatus*; *T. teguixin*; *T. rufescens*
temple turtle	*Heiremys anandalei*
tentacled snake; fishing snake	*Erpeton tentaculum*
tessellated water snake	*Natrix t. tessellatus*
thorny devil lizard	*Heteropterex dilatata*
tic-polonga snake; Russell's viper	*Vipera russelli*
tiger rat snake	*Spilotes pullatus*
tiger rattlesnake	*Crotalus tigris*
tiger snake	*Notechis scutatus*
timber rattlesnake	*Crotalus horridus*
Timor python	*Python timorensis*
toad-headed turtles	*Phrynops* sp.
Tokay gecko	*Gekko gecko*
trans Pecos rat snake	*Elaphe subocularis*
tree lizards	*Urosaurus* sp.
tree snakes	*Abaetulla* (*Dryophis*) sp.; *Imantodes* sp.
tropical chicken (or rat) snake	*Spilotes p. pullatus*

tropical rattlesnake; cascavel	*Crotalus durissus cascabel*
tuatara	*Sphenodon punctatus*
twin-spotted rattlesnake	*Crotalus pricei*
"two headed" snake	*Anilius s. scytale*
two-striped garter snake	*Thamnophis couchi hammondi*
Uracoan rattlesnake	*Crotalus vegrandis*
urutu, wutu snake	*Bothrops alternatus*
Vietnamese wood turtle	*Geomyda spengleri*
vine snakes	*Oxybelis* sp.; *Thelotornis* sp. *Uromacer* sp.
"viper" boa	*Candoia asper*
vipers, old world	*Cerastes; Echis; Pseudocerastes; Vipera* sp., etc.
Wagler's pit viper	*Trimeresurus wagler*
wall gecko	*Tarentola borgetti*
wart snake	*Acrochordus javanicus; A. granulosus*
water dragon lizards	*Physignathus* sp.
water moccasin	*Agkistrodon piscivorus*
water monitor lizard	*Varanus salvator*
water python	*Liasis olivaceus*
water snakes	*Natrix; Nerodia* sp.; *Enhydris* sp.
western ribbon snake	*Thamnophis proximus*
whip-tailed lizard	*Cnemidophorus* sp.
white-lipped forest cobra	*Naja melanoleuca*
white-lipped pit viper	*Trimeresurus albolabris*
white-lipped python	*Liasis albertisi*
wolf snake	*Lycodon aulicus; Lycophidion capense; L. striatus*
wood turtle	*Clemmys insculpta*
worm snakes	*Typhlops; Leptotyphlops*
worm lizards	*Bipes biporus; Blanus* sp.
yellow bellied slide turtle	*Trachemys s. scripta*
yellow-footed (legged) tortoise	*Geochelone denticulata*
yellow-lipped (pine woods) snake	*Rhadinaea flavilata*
yellow rat snake	*Elaphe obsoleta quadrivittata*
zebra tailed lizard	*Callisaurus draconoides*

Appendix E
Species List Cross-Referenced by Scientific Name

Please note: where a particular species may be known by more than one common name, each name is listed.

Abaetulla (Dryophis) prassinus	Long-nosed tree snake
Acalyptophis peronii	Peroni's sea snake
Acanthodactylus pardalis	sand lizard (gecko)
Acanthrophis antarcticus	death adder
Acanthosaurus armatus	mountain horned lizard
Acrochordus granulosus	elephant trunk snake; "kurung"
Acrochordus javanicus;	wart snake; "karung"
Agama sp.	agama lizards
Agkistrodon bilineatus	Mexican moccasin, cantil
Agkistrodon blomhoffi	mamushi snake
Agkistrodon contortrix	copperhead snake
Agkistrodon intermedius	Amur viper
Agkistrodon piscivorus	water moccasin snake
Agkistrodon rhodosotma	Malaysian moccasin
Alligator mississippiensis	American alligator
Alligator sinensis	Chinese alligator
Alsophis vudii picticeps	West Indian racer
Amblyrhynchus cristatus	Galapagos Isl. marine iguana
Ameiva ameiva	jungle runner lizard; rainbowlizard
Amphibolurus barbatus	bearded lizard
Anguis fragilis	glass "snake" lizard, slow "worm" lizard
Anilius s. scytale	"two headed" snake
Anniella pulcra	legless lizard

Anolis sp.	anole lizards
Aipysurus duboisi	Dubois's sea snake
Aipysurus sp.	sea snakes
Arizona elegans	glossy snake
Aspidelaps lubricus	shield nosed "cobra"; Cape coral snake
Aspidites melanocephalus	black-headed python
Astrotia stokesii	Stokes' sea snake
Atheris bispidus	Rough-scaled tree viper
Atheris squamiger	bush viper
Azemiops feae	Fea's viper
Basiliscus sp.	basilisk lizards
Bipes biporus	two-footed worm lizard
Bitis arietans	puff adder
Bitis atropos	mountain adder
Bitis caudalis	horned adder
Bitis cornuta	many-horned adder
Bitis g. gabonica; B. g. rhinoceros	Gaboon viper
Bitis nasicornis	rhinoceros viper
Bitis peringueyi	dwarf sand adder
Blanus sp.	two-footed worm lizards
Boa constrictor ssp.	common red-tail boas
Boiga dendrophila	mangrove snake
Boiga irregularis	brown tree snake
Bothrops sp.	new world pit vipers other than *Agkistrodon, Bothriechis, Crotalus, Lachesis, Ophryacus, Porthidium,* or *Sistrurus*
Bothrops alternatus	urutu, wutu snake
Bothrops andianus asper	fer-de-lance snake
Bothriechis (Bothrops) sp.	palm vipers
Bothriechis (Bothrops) schlegeli	eye lash viper
Boulengerina annulata;	eastern water cobra
Brachylophus sp.	Fiji Island iguanas
Brooksia sp.	dwarf chameleons
Bungarus fasciatus	banded krait
Bungarus caeruleus sindanus	blue krait
Bungarus m. multicinctus	many-banded krait
Caiman, sp. *Melanosuchus niger*	caimans
Calabaria reinhardtii	Calabar burrowing python
Calliopistes maculatus	Chilean dwarf tegu lizard
Callisaurus draconoides	zebra-tail lizard
Calloselasma rhodosotma	Malayan pit viper
Candoia asper	"viper" boa
Candoia bibroni	Solomon Island ground boa
Candoia carinata	New Guinea tree boa

Carettochelys insculpta	Fly River turtle
Carretta carretta	loggerhead turtle, marine
Causus resimus	green night adder
Cemophora c. coccinea	scarlet snake
Cerastes c. cerastes	horned desert viper
Cerastes c. gasperetti	hornless desert viper
Cerastes vipera	hornless desert viper
Cerastes cerastes	desert horned viper
Chalcides sexlineatus	sand skink
Chalcides sp.	Australasian frugivorous skinks
Chamaeleo sp.	old world chameleons
Charina bottae	rubber boa
Chelodina longicollis, etc.	long-necked turtles
Chelonia mydas	green sea turtle
Chelonoides chilensis	Chaco tortoise
Chelydra serpentina	common snapping turtle
Chelys fimbriata	mata mata turtle
Chinemys reevesi	Reeve's turtle
Chionactes sp.	shovel-nosed snake
Chironius scurrulus	green headed tree snake
Chitra indica	Indian soft shell turtle
Chondrodactylus sp.	geckos
Chondropython viridis	green tree python
Chrysemys sp,	cooter turtles
Chrysemys picta ssp.	painted turtles
Chrysolopea sp.	parrot snakes, flying snakes
Clelia clelia	mussurana
Clemmys guttata	spotted turtle
Clemmys insculpta	wood turtle
Clemmys muhlenbergi	bog, or Muhlenberg's turtle
Clemmys sp.	pond turtles
Clonophis kirtlandi	Kirtland's snake
Cnemidophorus sp.	race runner, whiptail lizards
Coleonyx sp.	banded geckos
Coluber constrictor ssp.	blue, yellow-bellied racer snakes
Coniophanes imperalis	black-striped snake
Conolophus pallidus	Galapagos land iguana
Corallus canina	emerald tree boa
Cordylus sp.	armadillo lizards; girdled lizard
Coronella girondica	old world smooth crowned snake
Corucia zebrata	prehensile tail skink
Corythophanes sp.	helmeted lizard
Crocodylus sp.	crocodiles
Crotalus sp.	rattlesnakes
Crotalus adamanteus	eastern diamondback rattlesnake
Crotalus atriocaudatus	canebrake rattlesnake

Crotalus atrox	western diamondback rattlesnake
Crotalus cerastes	sidewinder rattlesnake
Crotalus durissus	tropical rattlesnake; cascavel, cascabel;
Crotalus enyo	Baja California rattlesnake
Crotalus horridus	timber rattlesnake
Crotalus intermedius gloydi	Oaxacan small-headed rattlesnake
Crotalus lepidus ssp.	rock rattlesnakes
Crotalus mitchelli pyrrhus	speckled rattlesnake
Crotalus molossus	black-tailed rattlesnake
Crotalus polystictus	lance-headed rattlesnake
Crotalus pricei	twin-spotted rattlesnake
Crotalus pusillus	dusky rattlesnake
Crotalus ruber	red diamond rattlesnake
Crotalus scutulatus	Mojave rattlesnake
Crotalus stejnegeri	long-tailed rattlesnake
Crotalus tigris	tiger rattlesnake
Crotalus unicolor	Aruba rattlesnake
Crotalus vegrandis	Uracoan rattlesnake
Crotalus viridis	prairie rattlesnake
Crotalus viridis cerberus	black rattlesnake
Crotalus viridis helleri	southern Pacific rattlesnake
Crotalus viridis lutosus	Great Basin rattlesnake
Crotalus viridis oreganus	northern Pacific rattlesnake
Crotalus willardi	ridge-nosed rattlesnakes
Crotaphopeltis botamboeia	herald snake; cat-eyed snake
Crotaphytus collaris ssp.	collared lizard
Ctenosaura pectinata	spiny-tailed iguana
Cuora flavomarginata	Asian box turtle; snake-eating turtle
Cyclagras gigas	Brazilian smooth snake; false water cobra
Cyclemys sp.	leaf turtles
Cyclura sp.	rhinoceros iguanas; ground iguanas
Cylindrophis rufus	red-tailed pipe snake
Cylindrophis maculatus	spotted pipe snake
Cyrtodactylus pulchellus	gecko
Dasypeltis atra	egg-eating snake
Dasypeltis scabra	egg-eating snake
Deinagkistrodon acutus	sharp-nosed viper
Deirochelys reticularia	chicken turtle
Dendrelaphis punctulatus	Australian tree snake
Dendroaspis sp.	mamba snakes
Dermatemys mawei	Central American river turtle
Dermochelys coriacea	leatherback turtle, marine
Didodon rufozonatum	Asian big-tooth snake
Dipsas indica, D. sp.	snail-eating snakes

Dipsosaurus dorsalis	desert iguana
Dispholidus typus	boomslang snake
Dracaena guianensis	caiman lizard
Draco volans	flying lizard
Drymobius m. margaritiferus	speckeled racer
Drymarchon corais ssp.	indigo snake
Drymoluber dichrous	forest racer
Echis carinatus ssp.	rough (saw) scaled sand vipers
Echis sp.	sand vipers
Egernia sp.	frugivorous spiny-tail skinks
Elaphe sp.	rat (or chicken) snakes
Elaphe bimaculata	Chinese twin-spotted rat snake
Elaphe carinata	Chinese king rat snake
Elaphe climacophora	Japanese rat snake
Elaphe erythura	Philippine rat snake
Elaphe flavirufa	Mexican ratsnake
Elaphe flavolineata	New Guinea rat snake; copperhead rat snake
Elaphe g. guttata	red rat snake; corn snake
Elaphe helena	Indian rat snake
Elaphe l. longissima	Aesculapian snake
Elaphe moellendorffi	Moellendorff's rat snake
Elaphe obsoleta	black rat snake
Elaphe obsoleta emoryi	Great Plains rat snake
Elaphe obsoleta quadrivittata	yellow rat snake
Elaphe obsoleta spilotes	gray rat snake
Elaphe radiata	radiated rat snake
Elaphe rufodorsata	Chinese lined rat snake
Elaphe schrenckii	Russian rat snake
Elaphe subocularis	Trans Pecos rat snake
Elaphe taeniua friezi	Chinese rat snake; Asian beauty snake
Elaphe triaspis intermedia	green ratsnake
Elaphe vulpina	fox snake
Elgaria sp. (Gerrhonotus)	alligator lizards
Elysea novaguineae	New Guinea side neck turtle
Emydocephalus annulatus	Annulated sea snake
Emydoidea blandingi	Blanding's turtle
Enhydris plumbea	Indonesian green water snake; smooth scaled water snake
Epicrates angulifer	Cuban boa
Epicrates cenchria	Brazilian rainbow boa
Eremias grammica	dwarf sand lizard
Eretmochelys imbricata	hawksbill marine turtle
Ergernia sp.	frugivorous skinks
Eridiphas sp.	night snakes

Eristocophis macmahoni	leaf-nosed viper
Erpeton tentaculum	tentacled snake; fishing snake
Erythrolamprus bizona;	false coral snake
Eryx sp.	sand boas
Eublepharus macularius	leopard gecko
Eumeces sp.	insectivorous skinks
Eunectes murinus	anaconda
Eunectes notaeus	yellow anaconda
Exiliboa, Tropidophis, Ungaliophis	dwarf Caribbean boas
Farancia abacura	mud snake; horn snake,
Farancia e. erythrogramma	rainbow snake
Ficimia streckeri	hook-nosed snake
Gallotia stelini	Canary Island giant lacerta
Gavialis gangeticus	gavial
Gehyra sp.	geckos
Gekko gecko	Tokay gecko
Geochelone carbonaria	red-footed tortoise
Geochelone denticulata	yellow footed tortoise
Geochelone elephantopus ssp.	Galapagos tortoises
Geochelone elongata	elongated tortoise
Geochelone emys	Burmese mountain tortoise
Geochelone gigantea	Aldabra tortoise
Geochelone impressa	impressed tortoise
Geochelone pardalis ssp.	leopard tortoises
Geochelone sulcata	spurred turtoise
Geomyda spengleri	Vietnamese wood turtle
Geomyda spinosa	spiny turtle
Gerrhonotus sp.; *Elgaria* sp.	alligator lizards
Gerrhosaurus sp.	plated lizards
Glyalopion sp.	hook-nosed snakes
Gonatodes sp.	geckos
Gongylophis conicus	rough-scaled sand boa
Gonyosoma oxycephala	Asian rat snake; mangrove rat snake
Gopherus polyphemus	gopher tortoise
Gopherus (Xerobates) agassizi	desert tortoise
Gopherus (Xerobates) Berlandieri	Texas tortoise
Gopherus (Xerobates) flavo-marginata	Bolson tortoise
Graptemys geographica ssp.	map turtles
Graptemys oculifera	ringed sawback turtle
Graptemys pseudogeographica	false map turtle
Haemorrhois ravergieri	Ravergier's racer
Hardella thurgi	Brahminy River turtle
Heiremys anandalei	temple turtle
Heloderma horridum	Mexican beaded lizard
Heloderma suspectum	Gila monster lizard

Hemachatus hemachatus	Ringhal's cobra
Hemidactylus garnoti	house gecko
Hemitheconyx caudicinctus	fat-tailed gecko
Heterodon nasicus	western hog-nose snake
Heterodon platyrhinos ssp.	eastern hog-nose snakes
Heterodon simus	southern hog nose snake
Heteropterex dilatata	thorny devil lizard
Hydrophis sp.	sea snakes
Holbrookia sp.; *Cophosaurus* sp.	earless lizards
Hydrosaurus sp.	sail-tail; sailfin lizard
Hypsiglena sp.	night snake
Iguana iguana	common green iguana
Imantodes cenchoa	blunt-headed tree snake
Kinexys sp.	hingeback tortoise
Kinosternon sp.	mud turtle
Lacerta sp.	lacerta lizards
Lachesis muta	bushmaster snake
Lampropeltis calligaster	prairie kingsnake
Lampropeltis getulus californiae	California kingsnake
Lampropeltis g. getulus	eastern kingsnake
Lampropeltis getulus floridana	Florida kingsnake
Lampropeltis getulus holbrooki	speckled kingsnake
Lampropeltis getulus splendida	desert kingsnake
Lampropeltis mexicana alterna	gray-banded kingsnake
Lampropeltis mexicana greeri	Greer's kingsnake
Lampropeltis triangulum elapsoides	scarlet kingsnake
Lampropeltis triangulum	milksnake
Lampropeltis triangulum annulata	Mexican milksnake
Lampropeltis triangulum syspila	red milksnake
Lamprophis fulginosus	African house snake
Lapemis harwicki	Hardwick's sea snake
Laticauda sp.	Sea snakes
Leimadophis poecilogyrus	Argentine green speckled snake
Leiocephalus carinatus	curly-tailed lizard
Lepidochelys olivacea	olive ridley marine turtle
Leptodeira septentrionalis	cat-eyed snake
Leptophis ahaetulla; *L. mexicana*	parrot snakes
Leptotyphlops sp.	worm snakes
Lerista sp.	insectivorous skinks with reduced limbs
Liasis albertisi	white-lipped python
Liasis amethystinus	amethystine python
Liasis boa	Bismark ringed python
Liasis boeleni	Boelen's python
Liasis fuscus	brown water python

Liasis olivaceus	water python
Lichanura trivirgata	rosy boa
Lioheterodon madagascarensis	Malagasy giant hognose snake
Liolaemus sp.	Latin American swift lizards
Liopeltis sp.	smooth green snakes
Lycodon aulicus	wolf snake
Lycophidion capense; L. striatus	wolf snakes
Lystrophis sp.	South American hog-nosed snake
Mabuya sp.	frugivorous skinks
Macrochelys temminckii	alligator snapping turtle
Macropisthodon rudis	false habu snake
Malaclemys terrapin	diamondback terrapin
Malacochersus tornei	pancake tortoise
Malocochersus procteri	pancake tortoise
Malopon m. monspessulanus	Montpellier snake
Masticophis flagellum ssp.	coachwhip snakes
Maticora bivirgata flaviceps	Malayan long-glanded coral snake
Melanosuchus niger	black caiman
Micrurus sp.	eastern coral snakes
Micuroides fulvius	Arizona coral snake
Moloch horridus	moloch lizard
Morelia childreni	Children's python
Morelia spilotes variegata	carpet python
Naja melanoleuca	white-lipped forest cobra
Naja naja sputatrix	spitting cobra
Naja sp.	cobras
Natrix; Nerodia	water snakes
Natrix t. tessellatus	Dice snake; tessellated water snake
Notechis scutatus	tiger snake
Oligodon formosanus	Taiwan kukri snake
Opheodrys major	Asian smooth green snake
Opheodrys vernalis	rough green snake
Ophiophagus hannah	king cobra
Ophiosaurus apodus	sheltopusik legless lizard
Ophiosaurus ventralis	glass "snake" lizard
Ophryacus undulatus	undulated pit viper
Oxybelis sp.	vine snakes
Oxyuranus scutellatus	taipan snake
Pelamis sp.	sea snakes
Pelomedusa subrufa	African helmeted turtle
Pelusios sp.	East African side-neck turtles
Petrosaurus mearnsi	banded rock lizard
Phelsuma quadriocellata	four-eyed gecko
Phelsuma serratocauda	leaf-tailed gecko
Phrynocephalus mystaceus	Russian gargoyle lizard
Phrynops sp.	toad-headed turtles

Phrynops hilarii	Argentine side-neck turtle
Phyllodactylus sp.	leaf-toed geckos
Phyllorhynchus decurtatus	leaf-nosed snake
Physignathus sp.	water dragon lizards
Pituophis melanoleucus affinis	Sonoran gopher snake
Pituophis m. deppei	Mexican bullsnake
Pituophis melanoleucus deserticola	Great Plains gopher snake
Pituophis m. lodingi	black pine snake
Pituophis m. melanoleucus	northern pine snake
Pituophis m. mugitus	southern pine snake
Pituophis m. ruthveni	Louisiana pine snake
Pituophis m. sayi	bullsnake
Pituophis m. vertbralis	Baja bullsnake
Platemys megacephalum	big-headed turtle
Platemys spixii	spiny neck turtle
Podarcis sicula	ruin lizard
Porthidium (Bothrops) barbouri	Barbour's pit viper
Porthidium mummifer mexicanum	jumping viper
Psammophis sp.	sand snakes
Pseudapsis c. cana	mole snake, African
Pseudocerastes	false sand viper
Pseudohaje goldii	Gold's tree cobra
Pseutes poecilonotus polylepis;	mahogany rat snake,
Pseutes sulphureus	puffing snakes
Ptychozoon kuhli	flying gecko
Ptyodactylus hasselquistii	fat-footed gecko
Python anchietae	Angolan python
Python curtus	blood python
Python molurus	Indian rock python
Python molurus bivittatus	Burmese python
Python olivaceaus	olive python
Python regius	ball (regal) python
Python reticulatus	reticulated python
Python sebae	African rock python
Python timorensis	Timor python
Pytodactylus hasselquistii	fan-footed gecko
Regina septemvittata	queen snake
Regina sp.	crawfish snakes; swamp snakes
Rhabdophis subminiatus	Red-necked keelback snake
Rhadinaea flavilata	yellow-lipped (pine woods) snake
Rhamphiohis multimaculatus	African beaked snake
Rhamphotyphlops sp.	blind snake
Rhinoclemmys punctularia sp.	South American wood turtles
Sacalia bealei	four-eyed turtle
Salvadora hexalepis	patch-nosed snake
Sanzinia madagascariensis	Malagasy tree boa

Sauromalus obesus; *S. varius*	chuckwalla lizard
Scaphiodontophis annulatus bondurensis	neck-banded snake
Sceloporus magister	spiny lizard
Sceloporus sp.	fence lizard, swift, sage brush lizards
Scincella sp.	skink
Scincus scincus	sand "fish" skink
Seibenrockiella crassicollis	black pond turtle
Seminatrix p. pygaea	black swamp snake
Sistrusus catenatus ssp.	massasauga rattlesnake
Sistrurus miliarius	pygmy rattlesnake
Sistrurus ravus	Mexican massasauga snakes
Spalerosophis diadema cliffordi	diadem snake
Sphaerodactylus sp.	geckos
Sphenodon punctatus	tuatara
Spilotes pullatus	tropical "tiger" chicken or rat snake
Sternotherus odoratus	"stinkpot" turtle
Sternotherus sp.	musk turtle
Stilosoma extenuatum	sharp-tailed snake
Storeria dekayi	DeKay's snake
Storeria occipitomaculata	red-bellied snake
Tantilla sp.	black-headed snake; crowned snake; flat-headed snake
Tarentola borgetti	wall gecko
Tarentola sp.	geckos
Teratoscincus S. scincus	frog-eyed gecko
Terrapene sp.	box turtles
Testudo elegans	star tortoise
Testudo graeca	Greek tortoise
Testudo hermanni	Hermann's tortoise
Testudo kleinmanni	Egyptian tortoise
Testudo radiata	radiated tortoise
Thamnophis sp.	garter snakes, ribbon snakes
Thamnophis butleri	Butler's garter snake
Thamnophis couchi hammondi	two-striped garter snake
Thamnophis marcianus	checkered gartersnake
Thamnophis proximus	western ribbon snake
Thamnophis sirtalis	common garter snake
Thamnophis s. sackeni	peninsula ribbon snake
Thamnophis s. sauritus	eastern ribbon snake
Thamnophis s. similis	blue-striped garter snake
Thamnophis s. tetrataenia	San Francisco garter snake
Thasops jacksoni	black tree snake
Thelotornis kirtlandii	bird snake
Tiliqua sp.	blue-tongued skinks

Tomistoma schlegeli	false gavial
Trachemys s. scripta	yellow-bellied slider turtle
Trachemys scripta elegans	red-eared slider turtle
Trachyboa sp.	rough-skinned dwarf boa
Trachydosaurus sp.	old world frugivorous skinks
Trimeresurus albolabris	white-lipped pit viper
Trimeresurus flavoviridis	habu snake
Trimeresurus gramineus	Indian tree viper
Trimeresurus kanburiensis	Kanburian pit viper
Trimeresurus okinavensis	himehabu snake
Trimeresurus popeorum	Pope's tree viper
Trimeresurus purpureomaculatus	mangrove pit viper
Trimeresurus stejnegeri	Chinese tree viper
Trimeresurus sumatranus	Sumatran tree viper
Trimeresurus trigonocephalus	Sri Lanken pit viper
Trimeresurus wagleri	Wagler's pit viper
Trimeresurus wirotii	Siamese palm viper; Wirot's pit viper
Trimorphodon biscutatus	lyre snake
Trionyx sp.	soft-shelled turtles
Tropidoclonium lineatum	lined snake
Tropidodipsas sartori	snail eating snake, terrestrial
Tropidophis sp.	dwarf tree boa
Tupinambis nigropunctatus;	black and white/yellow tegu lizard
Tupinambis rufescens	red tegu lizard
Tupinambis teguixin	tegu lizard
Typhlops sp.	worm snakes
Uma sp.	fringe-toed lizards
Uromacer sp.	vine snakes
Uromastix acanthinurus;	Bell's dab lizard
Uromastix aegypticus	spiny-tailed lizard
Urosaurus graciosus	long-tailed brush lizard; tree lizard
Uta palmeri	side-blotched lizard
Uta stansburiana	side-blotched lizard
Varanus beccari	black tree monitor lizard
Varanus exanthematicus	savannah monitor lizard
Varanus grayi	Gray's monitor lizard
Varanus indicus	mangrove monitor lizard
Varanus karlschmidti	peach throat monitor lizard
Varanus niloticus	Nile monitor lizard
Varanus olivaceus (V. grayi)	Gray's monitor lizard
Varanus praesinus	green tree monitor lizard
Varanus rudicollis	rough-necked monitor lizard
Varanus salvadori	crocodile monitor lizard
Varanus salvator	water monitor lizard
Vermicella annulata	bandy-bandy snake

Vipera sp.	old world vipers
Vipera ammodytes	long-nosed viper; sand adder
Vipera aspis	asp
Vipera berus	European viper; adder; crossed adder; kreuzotter
Vipera kaznakovi	Kaznakov's viper
Vipera lebetina	spp. blunt-nosed vipers
Vipera orsinii	Orsini's viper
Vipera palaestinae	Palestinian viper
Vipera russelli	Russell's viper; tic-polonga
Virginia striatula	rough earth snake
Virginia valeriae	smooth earth snake
Walterinnesia aegyptia	Sinai Desert cobra
Xantusia sp.	night lizards
Xenopeltis unicolor	sunbeam snake
Zaocys dhumnades	keeled ratsnake
Zonosaurus sp.	plated lizards

Index of Taxonomic Names

Acalyptophis peronii, diet 12
Acanthrophis antarticus, diet 8
Acanthosaurus armatus, diet 24
Acheta domestica, culture 100
Acrochordus sp., diet 9
Agama sp., diet 20
Agkistrodon bilineatus, diet 7, 16
Agkistrodon contortrix, diet 7, 16
Agkistrodon piscivorus, diet 13, 16
Ahaetulla sp., diet 16
Aipysurus sp., diet 12
Aldabrachelys elephantina, diet 31
Alligator mississippiensis, diet 40; hypoglycemia 67; hypovitaminosis-K 66
Alligator sinensis, diet 40
Alsophis sp., diet 14
Amblyrhynchus sp., diet 23, 30
Ameiva ameiva, diet 23
Amphibolurus barbatus, diet 20
Anguis fragilis, diet 25
Anilius scytale, diet 9
Anniella pulcra, diet 23
Anolis sp., diet 20
Arizona elegans, diet 10
Aspidelaps lubricus, diet 7
Aspidites sp., diet 6
Astrotia stokesii, diet 12
Atheris sp., diet 16
Azemiops feae, diet 16

Bacillus thuringiensis 100,103
Baculovirus, infection of meal beetle larvae 103
Basiliscus sp., diet 20
Bipes biporus, diet 27
Bitis sp., diet 16
Blanus sp., diet 27
Boa sp., diet 6
Boiga irregularis, diet 7
Boiga dendrophila, diet 7, 11
Bombyx mori, culture 93
Bothrops sp., diet 13, 16
Bothriechis sp., diet 13, 16
Boulengerina annulata, diet 7
Brachylophus sp., diet 22
Brooksia sp., diet 20
Bungarus sp., diet 11

Caiman sp., diet 40
Calabaria reinhardtii, diet 6
Callisaurus draconoides, diet 27
Calloselasma rhodosotma, diet 13, 16
Candoia sp., diet 6
Caretta caretta, diet 35; silicosis from eating coral 81
Carettachelys sp., diet 33
Causus sp., diet 16
Cemophora sp., diet 15
Cerastes sp., diet 16

Chalcides sp., diet 25
Chamaesura anwina, diet 25
Chameleo sp., diets 20
Charina bottae, diet 6
Chelodina longicollis, diet 35, 37
Chelonia mydas, diet 34
Chelonoidis chilensis, diet 32
Chelydra serpentina, diet 37
Chelys fimbriata, diet 35
Chionactes sp., diet 15
Chironius sp., diet 14
Chlamydosaurus kingii, diet 21
Chondrodactylus sp., diet 22
Chondropython sp., diet 6
Chrysemys sp., diet 32,36
Chrysolopea sp., diet 13
Clelia clelia, diet 12
Clemmys sp., diet 32, 35, 36, 37
Clonophis kirtlandi, diet 11
Cnemidophorous sp., diet 27
Coleonyx sp., diet 22
Coluber sp., diet 14
Coniophanes imperalis, diet 6
Conolophus pallidus, diet 23
Contia tennuis, diet 15
Cophotus sp., diet 21
Corallus sp., diet 6
Cordylus sp., diet 20, 26
Coronella girondica, diet 8
Corucia zebrata, diet 25
Corythophanes sp., diet 22
Cricosaura typica, diet 24
Crocodylus sp., diet 40
Crotalus sp., diet 13, 14
Crotaphopeltis botamboeia, diet 10
Crotaphytus, sp., diet 20
Ctenosaura pectinata, diet 26
Cuora sp., diet 31
Cyclagras sp., diet 9
Cyclemys sp., diet 34
Cyclura sp., diet 23
Cylindrophis sp., diet 13

Dasypeltis scabra, diet 9, 63
Deinagkistrodon, diet 13, 16
Deirochelys reticularia, diet 32

Delma fraseri, diet 25
Dendrelaphis punctulatus, diet 5
Dendraspis sp., diet 11
Dermatemys mawei, diet 37
Dermochelys coriacea, diet 34
Diodophis sp., diet 14
Dipsas indica, diet 15
Dipsosaurus dorsalis, diet 22
Dispholidus typus, diet 6
Dracaena guianaensis, diet 19, 20, 30
Draco volans, diet 21
Drosophila melanogaster, culture 103
Drymarchon sp., diet 8, 10
Drymobius margaritiferus, diet 14
Drymoluber dichrous, diet 14

Echis sp, diet 16; water requirements 41
Egernia sp., diet 25
Elaphe sp., diet 5, 8, 9, 14
Elgaria sp., diet 20
Elysea novaguineae, diet 36
Emydocephalus sp., diet 12
Emydoidea blandingi, diet 31
Enhydris sp., diet 12
Enhyalius catenatus, diet 20
Epicrates sp., diet 6
Eremias grammica, diet 21
Eretmochelys imbricata, diet 34
Eridiphas sp., diet 13
Eristocophis, diet 16
Erpeton tentaculum, diet 16
Erythrolampus bizona, diet 9
Eryx sp., diet 6
Eublepharus sp., diet 22
Eumeces sp., diet 25
Eunectes sp., diet 5
Exiliboa sp., diet 6

Farancia abacura, diet 12
Farancia erythrogramma, diet 14
Ficimia streckeri, diet 10
Fijivirus, infection of silk moth larvae 98

Index of Taxonomic Names

Galleria mellonella, culture of 105
Gallotia sp., diet 23
Gambelia sp., diet 23
Gavialis gangeticus, diet 40
Gehyra sp., diet 22
Gekko sp., diet 22
Geochelone carbonaria, diet 36
Geochelone denticulata, diet 38
Geochelone elegans, diet 37
Geochelone elephantopus ssp., diet 33
Geochelone elongata, diet 33
Geochelone (Testudo) gigantea, diet 31
Geochelone impressa, diet 34
Geochelone pardalis, diet 35
Geochelone radiata, diet 36
Geochelone sulcata, diet 37
Geomyda sp., diet 37
Gerrhonotus sp., diet 20
Gerrhosaurus sp., diet 24
Glyalopion sp., diet 10
Gonatodes sp., diet 22
Gopherus (Xerobates) sp., diet 32, 33
Gonatodes sp., diet 22
Gongylophis sp., diet 6
Gonyosoma oxycephala, diet 5
Graptemys oculifera, diet 33, 35, 37
Graptemys pseudogeographica, diet 33
Gryllus domesticus, culture 100

Haemorrhois ravergieri, diet 14
Hardella thurgi, diet 32
Heloderma sp., diets 20, 22; appetite stimulation 49
Hemachatus sp., diet 7
Hemidactylus sp., diet 22
Hemitheconyx sp. diet 22
Heterodon sp., diet 10
Heteropterex dilatata, diet 26
Hibiscus as a living forage plant in cages 28

Holbrookia sp., diet 21
Hydrophis sp., diet 12
Hydrosaurus amboinensis, diet 24
Hypsiglena sp, diet 13

Iguana iguana, diet 19, 22, 28
Imantodes cenchoa, diet 6

Japalura sp., diet 24

Kinexys sp., diet 34
Kinoststernon sp., diet 35

Lacerta sp., diet 23
Lachesis muta, diet 7, 13, 16
Lactobacillus acidophilus 62
Lactobacillus planarum 62
Lamprophis sp., diet 5
Lampropeltis sp., diets 10, 12
Lapemis sp., diet 12
Laticauda sp., diet 12
Leimadophis poecilogyrus, diet 5, 15
Leiocephalus carinatus 21
Lepidochelys olivacea, diet 36
Leptodeira septentrionalis, diet 7
Leptophis sp., diet 13
Leptotyphlops sp., diet 16
Lerista sp., diet 25
Lialis burtonis, diet 25
Liasis sp., diet 6
Lichanura sp., diet 6
Leioheterodon madagascariensis, diet 10
Liolaemus sp., diet 26
Liopeltis sp., diet 10
Lumbricus terrestris, culture 98
Lycodon sp., diet 16
Lycophidion sp., diet 16
Lyriocephalus scutatus, diet 23
Lystrophis sp., diet 10

Mabuya sp., diet 25
Macrochelys temminckii, diet 31
Macropisthodon rudis, diet 9
Malaclemys terrapin, diet 33

Malacochersus tornieri and *M. procteri*, diet 36
Malpolon monspessulanus, diet 12
Masticophis flagellum; diet 7, 16
Maticora bivirgata flaviceps, diet 11
Melanosuchus niger, diet 40
Micruroides sp. diet 8
Micrurus sp., diet 8
Moloch horridus, diet 23; water uptake 42
Monilia anisopliae 103
Morelia sp., diet

Naja sp., diet 7
Natrix sp., diet 15
Neoseps reynoldsi, diet 25
Nerodia sp., diet 16
Nosema bombycis, infection of silk moth larvae 98
Notechis scutatus, diet 16

Opheodrys sp., diet 10
Ophiophagus hannah, diet 11
Ophiosaurus sp., diet 22, 25
Ophryacus undulatus, diet 13
Oxybelis sp., diet 16
Oxyuranus scutellatus, diet 15

Pelamis, diet 12
Pelomedusa sp., diet 31
Pelusios sp., diet 31
Petrosaurus mearnsi, diet 24
Phelsuma sp., diet 22
Phrynocephalus mystaceus, diet 24
Phrynops sp., diet 31, 37
Phrynosoma sp., diet 22; water uptake 42
Phyllodactylus sp., diet 22
Phyllorhynchus decurtatus, diet 11
Physignathus sp., diet 27
Pituophis, diets 7, 10, 13
Platemys macrocephala, diet 31
Podarcis sicula, diet 24
Porthidium (*Bothrops*) sp., diet 13, 16

Psammodromus sp., diet 24
Psammophis sp., diet 15
Pseudaspis sp., diet 12
Pseudocerastes sp., diet 16
Pseudohaje goldii, diet 7
Pseutes sp., diet 12
Ptychozoon kuhli, diet 22
Ptyodactylus sp., diet 22
Pyrausta as a prey species for lizards 30
Python, sp., diet 6
Python regius prolonged fasts in, see anorexia 50

Regina sp., diet 8, 14
Reovirus, infection of insects 98
Rhabdophis subminiatus, diet 14
Rhadinaea flavilata, diet 16
Rhamphiophis multimaculatus, diet 5
Rhamphotyphlops, sp., diet 6
Rhinocheilus lecontei, diet 11

Saccalia bealei, diet 33
Sauromalus obesus; *S. varius*, diet 21
Salvadora sp., diet 13
Scaphiodontophis annulatus bondurensis, diet 12
Sceloporus sp., diet 25, 26
Serratia marcescens, infection of insects 103
Scincella lateralis, diet 25
Scincus sp., diet 25
Seibonrockiellaa crassicollis, diet 31
Seminatrix pygaea, diet 6
Sibon sp., diet 15
Sistrurus sp., diet 13, 14
Sonora semiannulata, diet 10
Spalerosophis diadema cliffordi, diet 8
Sphaerodactylus sp., diet 22
Sphenodon punctatus, diet 41
Spilotes pullatus, diet 16
Sternotherus sp., diet 36

Stilosoma extenuatum, diet 15
Storeria dekayi, diet 8
Storeeria occipitomaculata 14
Streptococcus faecium, commercial culture for intestinal flora restoration 62

Tantilla sp. diet 6, 8, 9
Tarentola sp., diet 22
Tenebrio molitor, culture 101
Teratoscinctus scincus, diet 25
Terrapene sp., diet 32
Testudo elegans, diet 37
Testudo graeca, diet 33
Testudo hermanni, diet 34
Testudo kleinmanni, diet 33
Thamnophis sp., diets 9, 14
Thasops jacksoni, diet 6
Thelotornis kirtlandii, diet 5
Tiliqua sp., diet 25
Tomistoma schlegeli, diet 40
Trachemys sp., diet 36
Trachyboa sp., diet 6
Trachydosaurus sp., diet 25
Trimeresurus sp., diet 13, 16
Trimorphodon biscutatus, diet 11
Trionyx sp., diet 37
Tropidoclonium lineatum, diet 11
Tropidodipsas sartori, diet 15
Tropidophis sp., diet 6
Tupinambis sp., diet 26, 63
Typhlops sp., diet 16

Uma sp., diet 21
Ungaliophis, sp., diet 6
Urocentron sp., diet 26
Uromacer sp., diet 16
Uromastix sp., diet 26
Urosaurus graciosus, diet 23, 26
Uta sp. diet 25

Varanus sp., diet 23, 63
Varanus grayi, diet 23, 29
Vermicella anulata, diet 5
Vipera sp., diet 16
Virginia sp., diet 9

Walterninnesia aegyptia, diet 7

Xantusia sp., diet 24
Xenodon sp., diet 10
Xenopeltis unicolor, diet 15

Zaocys dhumnades, diet 10
Zebrina pendula, as a living forage plant in cages 28
Zonosaurus sp., diet 22

General Subject Index

alfalfa pellets, as litter 68, 82
ammonium ion excretion 41
anorexia 48
appetite stimulation 44
articular and periarticular gout 41
articular and periarticular pseudo-gout 72
ascorbic acid synthesis, in snakes 63
autoovophagy 80
avidin, in biotin deficiency 62

beans, lentils, peas and seed sprouts, instructions for growing 89–91
Bene-Bac 62
beneficial intestinal microflora 62
biotin deficiency 62
bloating 72
bouyancy control 81

Calcimar 57
calcitonin, for treatment of hypercalcemia 57
calcium 50, 56
calcium, excess in diet 56
calcium metabolism 50
cannibalism 18, 79
carbohydrate fermentation 72
cardiomyopathy, dietary 64

carotene; carotenoids, natural sources; see vitamin A
chelonian diets 30–39
chitinous exoskeletons, of invertebrates 21
chloride excretion see sodium chloride
cobalt 58
commercial products suitable for feeding turtles 109–110
constipation 68
copraphagy 80
crocodilian diets 40
culture of invertebrates as prey 92

dehydration; see water requirements; gout induction 41
dehydration, treatment of 41
dermatophagy 79
diarrhea 71
diet, chelonian 30–39
diet, crocodilian 40
diet, lizard 19–30
diet, snake 4–19
diet, tuatara 41
dioctyl sodium sulfusuccinate 69

effects of feeding excessive amounts of dog, cat, and monkey chow 56

excessive dietary calcium 56
extra-renal excretion of electrolytes 42

feeding periodicity 44
fermented fruit, intoxication from ingestion of 78
fibrous osteodystrophy; see metabolic bone disease
Flagyl, for appetite stimulation: see metronidizol
food values 86–87
forced feeding 44, 50
frequency of feeding 44
frozen fish, thiaminase content 109
fruit flies, culture of 103

gastrointestinal silicosis, in marine turtles 81
gecko, diets 22
geophagy 81
goiters, diet induced 66
gout 41
growing green fodder 83

hibernation 91
high-fiber herbivore diet 91
hydroponic cultivation of grasses 88
hypercalcemia, treatment for 57
hyperkeratosis, in hypovitaminosis A 58
hyperparathyroidism 55
hypervitaminosis A, iatrogenic 18, 60
hypoglycemia, in captive crocodilians 67
hypovitaminosis A; see vitamin A deficiency 58
hypovitaminosis A, treatment of 59
hypovitaminosis K 66

iguana, diets 19, 22, 28, 91
improving dietary quality of prey species 95, 102
inanition 48
inappetance; see anorexia

intestinal microflora, restoration of 62
iodine 66
iron 58

kidneys 41

lipid-soluble vitamins 58, 50–53, 64, 66
lithophagy 81
lizard diets 19–30

maceration of ingesta 81
magnesium 58
manganese 58
mealworms, culture of 101
mealworms, chemical composition of 102
mealworms, improving nutritional quality of 102
metabolic bone disease 50
metabolic bone disease, treatment of 53
metronidazol 48
mice, mineral composition of 52
mineral imbalances 50
monitor lizard, diet 23, 29, 63
mouse pups, mineral composition of 52
mulberry silk moth larvae, culture of 93
mulberry silk moth larvae, chemical composition of 97

nitrogen excretion 41
nutrition-related behavioral disorders 79

obesity 44
obstipation 68, 82
osteomalacia; see metabolic bone disease
overfeeding 44

parathyroid adenoma 55
parathyroid tumors 55
phosphorus 50

phosphorus metabolism 50
poisonous plants 73–78
potassium excretion 42, 44
potassium 42
primary hyperparathyroidism 55
Probiocin 62
pseudogout 72

ready sources of diets 83
regional herpetological
   societies 113–122
rehydration 42
renal-associated fibrous
   osteodystrophy 55
rickets 54

safe and nutritious plants 84–85
salt glands 42
selenium 66
simethicone 69, 72
skink, diet 25
snake diets 4–19
sodium chloride 42, 43
sodium excretion 42
sources of living and frozen reptile
   food 107
squamous metaplasia, in
   hypovitaminosis-A 58
starvation 48
stimulation of appetite 48

thiaminase 61

thiamine deficiency 61
toxic plants 73–78
trace minerals 58
tuatara, diet 41
tympany (bloating) 72

ultraviolet radiation 49; also see
   calcium metabolism
urates 41
uric acid excretion 41

visceral gout 41
vitamin A 58
vitamin B-1, thiamin 61
vitamin C; ascorbic acid 63
vitamin D; vitamin D-3
   deficiency 50, 53, 64;
   overdosage 18, 28
vitamin E 64
vitamin E deficiency 64
vitamin K, deficiency 66
vomiting, causes of 70
vomiting, differential diagnosis
   of 70

water requirements 41
wax moths, culture of 105
white muscle disease 65

xeroderma 60